LIEUTENANT OWEN WILLIAM STEELE OF
THE NEWFOUNDLAND REGIMENT

Lieutenant Owen W. Steele

Lieutenant Owen William Steele of the Newfoundland Regiment

Diary and Letters

EDITED BY
DAVID R. FACEY-CROWTHER

MCGILL-QUEEN'S UNIVERSITY PRESS
MONTREAL & KINGSTON · LONDON · ITHACA

© McGill-Queen's University Press 2002
ISBN 0-7735-2428-2
Legal deposit fourth quarter 2002
Bibliothèque nationale du Québec

Printed in Canada on acid-free paper that is 100% ancient forest free
(100% post-consumer recycled), processed chlorine free.

This book has been published with the help of grants from the
Humanities and Social Sciences Federation of Canada, using funds
provided by the Social Sciences and Humanities Research Council of
Canada, and the Publication Subvention Committee of Memorial
University of Newfoundland.

McGill-Queen's University Press acknowledges the financial support of
the Government of Canada through the Book Publishing Industry
Development Program (BPIDP) for its publishing activities. We also
acknowledge the support of the Canada Council for the Arts for our
publishing program.

National Library of Canada Cataloguing in Publication Data

Steele, Owen William
The diary and letters of Lieutenant Owen William Steele of the
Newfoundland Regiment / edited by David R. Facey-Crowther

Includes bibliographical references and index.
ISBN 0-7735-2428-2

1. Steele, Owen William. 2. World War, 1914–1918 – Personal
narratives, Newfoundland. 3. Great Britain, Army, Newfoundland
Regiment – Biography. I. Facey-Crowther, David R. (David Richard),
1938– . II. Title

D640.S718 2002   940.4'81718   C2002-902771-3

Typeset in 10/12 Sabon by True to Type

Contents

Acknowledgments

This publication would not have been possible without the generous assistance of a host of individuals, many former students of mine at Memorial University of Newfoundland who, over the years, worked on the text and annotations. In particular I want to thank Dean Menchions, Bill Cove, Gale Warren, Wendell Head, Jason Churchill, and Brad Anthony. I am especially indebted to James Steele for his work on the family history that forms part of this publication, for his tireless efforts in identifying individuals who appeared in the diary and letters, and for his careful reading of both the original and typescript for errors and omissions. His personal collection of photographs and mementoes belonging to both his father and uncle provided the illustrations for this book.

Preface

Shortly after the Steele family was informed of their son Owen's death they pressed the governor, Sir Walter Davidson, to obtain some of his personal effects connected with his military service, specifically his diary and his Brownie camera. The diary was returned to the family shortly thereafter and in March 1917, following submission to the censor, was bound and presented to the Standing Committee of the Newfoundland Patriotic Association for publication. S.O. Steele's intention was to give the entire proceeds of the sale of the book to the Newfoundland Patriotic Association. By then, however, a narrative of the war history of the Newfoundland Regiment was already being prepared, based in considerable measure on the Steele diary. In consequence, the diary was not published – but neither was the proposed narrative. The official account of the exploits of the Royal Newfoundland Regiment in the Great War had to wait another fifty years until Lt. Col. G.W.L. Nicholson's magisterial *The Fighting Newfoundlander* was finally published.

The typescript for the bound copy had been prepared by Owen's sister Ella and included both the diary and letters sent to the family over the course of Owen's service with the regiment. The original letters, regrettably, have been lost. Ella's transcriptions appear in the typescript but to what extent they were edited is not known. All one can say is that the writing style seems consistent with the diary. Ella probably eliminated family detail that she felt was not relevant to the context of the diary entries. Only in one instance did she alter the text of the diary: when she thought that Owen's description of the regiment's reaction to Hadow's strict discipline in Egypt cast the unit in a bad light.

The original diary and a typescript copy of both the letters and the diary have remained in the possession of the Steele family. James R. Steele, Owen's nephew, whose father James had fought at Beaumont Hamel, provided the Centre for Newfoundland Studies Archives on the campus of Memorial University of Newfoundland with xerox copies of the originals. It was always Mr Steele's intention to have the diary and letters published and my interest in bringing this important source on Newfoundland's war experience to a broader reading audience thus coincided with James Steele's wishes. I am grateful to him for allowing me the opportunity to edit the diary and letters. It is his intention and mine that this volume should be dedicated to both James and Owen, who shared so much in their early lives together.

A Family History
James H. Steele

My earliest memories are of living over the shop at 100 Water street with my parents, James R. and Amy (Stevenson) Steele, my brother, Owen, and my sister, Isabelle. The building was four storeys high and of brick construction. My brother and I shared a bedroom on the top floor, which was reached by climbing three flights of stairs, and it had a magnificent view of the Narrows. On the third floor there were two bedrooms, my parent's and my sister's, a kitchen, a dining room, and bathrooms. The "sitting room" was on the second floor. It was a large room, comfortably furnished and heated by a coal fireplace. It was the centre of our home, a place where we gathered Sunday evenings after dinner, to be entertained by Isabelle on the piano or to listen to our favourite gravelly recordings on the gramophone. When our parents had company we usually headed up to our rooms where we managed to entertain ourselves with little difficulty until bedtime.

The St John's summers are short and glorious and as small children we spent as much time as possible outdoors or in the country. Winters, on the other hand, usually meant long hours indoors in cold and draughty rooms where the only source of heat came from small coal fireplaces. Despite this, the winters never seemed to be confining. Perhaps it was because there was always so much to see and do. From the fourth storey windows we could look out at the Narrows and

S.O. Steele and
Sons Ltd, St. John's,
ca 1914

watch the sleek Banks schooners and sturdy coastal steamers heading
to and from the city's busy mercantile establishments. The office and
shop were on the ground floor with a packing room in the back and
a warehouse loading door that opened on Holloway Street. S.O.
Steele was one of the city's smaller firms, run by brothers James R.
and Victor J., and had supplied Newfoundland's outports with practical English tableware and earthenware since the turn of the century. It was a busy place with customers and goods coming and going
over the course of the business day. There was much there to interest
young curious eyes. Over the course of our childhood very little
changed.

I have in my possession a finely penned postcard from Devon, England, dated 5 May 1934, sent to me by my grandfather, Samuel Owen
Steele, the founder of the family firm. I was nine at the time. On it he
stated that just fifty-two years earlier he had landed in St John's.
Recently, when I started to trace the family history, I looked at a copy
of *The Evening Telegram* dated 5 May 1882 and found the notice with
details of his arrival in St John's.

St John's, city and harbour, ca 1900

The s.s. *Hibernian* – first direct mail boat for the season - arrived here from Liverpool and Queenstown between 6 and 7 o'clock this morning. Her Captain, as far as we are aware, reports nothing of unusual importance. The water seemed just as wet as ever, and "the wind blew where it listed," without evincing any special regard for the ancient craft. She brought a large quantity of freight for our dry-goods men and the following ladies and gentlemen came as passengers:-

Mrs. Dance, Mrs Gear, Mrs. Gibb, Miss Gibb, Miss Tobin, Messrs. Alexander, Bulwer, Byone, Dance, Gear, Pomeroy, Poole, Robinson, Rothwell, Schofield, Sharpe, Steele, Tobin, Master K. Gibb, and nine in steerage.

Samuel Owen Steele was born in Manchester on 23 January 1862, the first of Samuel and Mary Maria's seven children.[1] After a brief stint in the army, Samuel migrated to St John's, where he seems to have taken a job as a draper. He was an active member of the local Congregational church, the precursor of the present St David's Presbyterian Church. It was there that he met his future wife, Sarah Blanche Harris.

Sarah Harris was a native of Teignmouth, Devon, the eldest daughter of Mr and Mrs Richard Harris. At the age of eleven she was sent over to Newfoundland to live with her aunt and uncle, James H. and Hannah (Tucker) Martin. "I cried all the way over," she later confessed. Why someone so young should have been forced to leave home and family remains a mystery but the fact that the Martins had lost all four of their own children at early ages might provide an answer. She seems to have settled in well with her relatives and, like them, to have taken an active role in church affairs. For Sarah, as for Samuel, the church was the centre of their social life. They married on 29 May 1886.

The marriage marked an important change of fortune for the young Samuel. His wife's uncle (who was effectively his father-in-law) was already a successful St John's merchant. From humble beginnings as a wharfinger for Job Brothers, James Martin quickly established himself as an independent shopkeeper. By the mid 1860s he had a shop in downtown St John's, at 98-100 Water Street, that sold a "large and varied assortment" of china and earthenware to both city and outport customers. Eventually he bought out his former boss's business, Job's Hardware, and renamed it J.H. Martin & Co. This business became the foundation for a family business that in time included two nephews, William and Frank Martin, brought over from Devon to help manage the firm. By the mid 1870s James Martin was sufficiently well-established to expand his business interests to include rental property. He commissioned Devonshire architect James Southcott to build Devon Row, five fashionable townhouses on Plymouth Road, over-looking the harbour entrance. (Devon Row miraculously survived the Great Fire of 1892 and four of the houses have been recognized as des-ignated property by St John's Heritage Trust.)

In September 1886 Samuel Steele opened his own business, The Manchester House on 101 Water Street, an emporium that carried a large and varied stock of millinery, leatherware, and general dry goods.[2] The following year, during the annual August regatta on Quidi Vidi Lake, Steele distributed nearly a thousand Jubilee medals marking Queen Victoria's 50th anniversary and also advertising his store. Steele was also a participant in that year's regatta and was a member of the winning crew that won the President's Prize for the race sponsored by the City Boat Club. The other members of the winning crew, all con-nected with St John's commercial establishment, were W.R. Rennie, G.R. Bowring, T. Pittman, D. Browning, and T. Steer – which I know from the inscribed trophy, bought at an antique shop in 1976.

In July 1892 a great fire engulfed St John's, wiping out much of the city including a large part of its commercial life. Steele had moved to a new premises at 129 Water Street in September 1889 which had for-merly been occupied by Richard Harvey.[3] This new enterprise and Martin's Crockeryware and Hardware shops were among the victims of the fire. J.H. Martin & Co was quickly back on its feet with tem-porary quarters located at Seamen's House at 353 Duckworth. The

Steele family home, Trentham, 1909

crockeryware shop, under Hannah Martin's management, relocated to 31 Plymouth Road, Hoyles-town, until 1894 when new premises were constructed at 100 Water Street. Samuel Steele worked for nearly two years as agent for the American Furniture Company, offering a wide range of furniture and tapestries to a city eager to rebuild and refurbish after the fire. In 1894 he was ready to open his own furniture business, this time at the new Total Abstinence Hall on Duckworth Street, a popular spot for local theatre.

This new Steele enterprise lasted less than a year. On 10 December 1894 the Union and Commercial Banks in St. John's suspended payment, bringing on what was later to be called Newfoundland's "Black Monday." Bank notes were valueless and trade everywhere was at a standstill.[4] Some of the larger firms, such as Martin's Hardware and the Crockeryware Shop, were able to survive the crisis but small enterprises such as Steele's fell victim to the financial crisis. For the next five years Samuel worked as a clerk. Then in May 1899 Hannah Martin had a stroke. She had managed the Crockeryware Shop right up to the time of her death. Before the year was out her disheartened husband turned over the business to Samuel and retired to England.

Samuel and Blanche were then living in Mundy Pond, in the west end of St John's, in a farm house previously owned by Hugh Hamlyn. Christened "Trentham House," after Samuel's ancestral home in

England, it became home for the next twenty years to the Steele's bur-
geoning young family. Between 1887 and 1899 the Steeles produced ten
children including Owen William, the first, and my father, James Robert.[5]

The three oldest boys, Owen, James, and Herbert, grew up together
and, when not at school, kept busy either tending the farm or helping out
at the shop. All three were keen athletes. The distance from the house at
Mundy Pond to the shop at 100 Water Street was over 4 kms and across
a height of land 100 metres above sea level. Needless to say, the boys
became good walkers and in time champions in the twenty-mile walking
races at the Highland Games sponsored by the local St Andrew's Club.

Schooling for the Steele children was a mixture of private tutoring and
public school. I have in my possession an invoice from Louise E. Barnes
for tuition for "Master Owen" for the quarter ending 1 February 1894 for
$2.00, and another, from M.J. Dicks of Mundy Pond Road, dated 3
October 1894, for three months tuition for the quarter ending 28 September, for Master Owen, $2.00, and Master James, $2.00. Owen and
James both attended Methodist College for a short while. At the turn of
the century the three oldest boys were at the Church of England's Bishop
Feild College, then under its distinguished Yorkshire-born headmaster, Dr
William Blackall.

In April 1902 my father James, then only fourteen, was taken out of
school to work on the farm, much to his disgust. Apparently, he was
doing quite well at the time, having gained second place for proficiency
in mathematics that year. He later worked for a period in the family business and also in a large department store on Water Street before joining
the Newfoundland Regiment in January 1915.

Owen and Herbert eventually graduated from Bishop Feild. Both
excelled in penmanship, a skill Owen would later put to good use as a
diarist and letterwriter during the Great War. The following is taken from
the college yearbook *The Feildian* in 1901 and it indicates something of
the early character of this remarkable young man, then only fourteen.

Owen W. Steele entered the College in September 1899 and has passed the Pre-
lim and Interm Grades of C.H.E. with credit. During the first two years his
attendance was remarkable for its regularity. He is a boy of good intelligence
and of good moral conduct. He has started life in one of the offices on Water
Street. We wish him well.

On graduation Owen went to work full-time in the family business.
While his father built up and expanded the crockeryware shop, especially the wholesale part, Owen travelled in the outports as salesman for
the firm. There was stiff competition for business. Several St John's
shops specialized in crockeryware and all the major department stores
carried a full stock of items. There was very little money to spare. Prior

Owen Steele in the 1909 twenty mile walking race. Gordon V. Boone is on his right, Hugh A. LeMessurier looking over his left shoulder

Owen and James Steele, shown with the trophy for the twenty mile walking race, 1909

to the war efforts were made to take on other lines, such as wholesale groceries, but later dropped. The firm did expand its stock of hardware, including enamelled ware, cutlery, and aluminum and galvanized ware.

Although now out of school, the three older brothers continued to participate in local sports activities, especially the walking matches sponsored by the St Andrew's Club. To keep in shape they practised regularly, often early in the morning before work. Their efforts brought many distinctions. I have in my possession gold and silver medals won by my uncle and father and a 1909 photograph of the two of them standing in front of the twenty-mile race trophy (actually 20¾ miles) they won that year. The trophy plaque shows how often they had won this race.

ST. ANDREWS CLUB – 20 MILE WALKING MATCH

1907	OWEN W. STEELE	3 Hours 46 Min. 17 Sec.
1908	OWEN W. STEELE	3 Hours 27 Min. 29 Sec.
1909	OWEN W. STEELE	3 Hours 13 Min. 37 Sec
1909	JAMES R. STEELE	3 Hours 20 Min. 45 Sec.
1910	JAMES R. STEELE	3 Hours 20 Min 7 Sec.

ST. ANDREWS CLUB – 1 MILE WALK (TURF)

1913	JAMES R. STEELE	8 Minutes

The Steele family were very musical. Owen sang in the choir of the Congregational Church with Dick, Mary (Molly), and Frances. Owen was secretary of the Children's Festival organized to mark the coronation of King George V on 23 June 1911 and held on the grounds of Government House. *The Evening Telegram* reported that the festival was a great success with 5,000 children in attendance. Richard, the youngest brother, left school in 1911 and received his musical education at the Royal Academy of Music in London where he was awarded the bronze medal for piano.[6]

In the years before the Great War the Steele family seemed ready to settle down in their life's pursuits. Owen and Jim were already employed with the family firm. Richard, one of the youngest, eventually studied music in England but, although he did well as a student, never followed a musical career. Instead, he returned to Newfoundland to work in the family business during the war, setting out on his own in 1923 with a crockeryware shop on Water Street opposite the Court House.

On 4 August 1914 Great Britain went to war against Germany. The Newfoundland government offered to raise troops for land service abroad, an offer gladly accepted by His Majesty's government. Owen W. Steele was one of the first 500 to volunteer. He signed up on 13 September and embarked for England the following month on board the SS *Florizel*. His diary and letters are the subject of this book

Newfoundland was yet to brace for the tragedies that marked the Regiment's participation in the European conflict. The first wartime loss for the Steele family was not, however, the result of battle: Herbert returned

S.O. Steele and Sons Ltd, St. John's, ca 1914

from Baltimore with the dreaded disease tuberculosis. (We were told later that he had caught it while practising dentistry on the poor in Baltimore.) The family was devastated when he died a few months later, on 15 March. His funeral took place from the premises at 100 Water Street. Family mourning in those days was strictly observed over an extended period of time and respected by other members of the community. When the Benevolent Irish Society procession passed the premises on 17 March the band ceased playing and the procession halted out of respect. My grandfather was so touched that he wrote to the *Evening Telegram* thanking the BIS for its thoughtfulness: "It is such acts of fellow-feeling to those in trouble that make the whole world kin."

My father enlisted in the Newfoundland Regiment on 11 January 1915 and left with the second contingent the day after Herbert's funeral. He also kept a diary of his war experiences, although not nearly as extensive a one as Uncle Owen's. (Excerpts from his diary have been included as illustrative material in this book.) Lt James R. Steele was wounded a second time at Ypres on 26 March 1918. After convalescence, on 2 July 1918 he was drafted for special duty home to Newfoundland and put in charge of a contingent of thirty men to guard Bell Island from possible German attack by submarines. He was retired from active duty 10 March 1919 and returned to the shop.

Ella Steele (left) with sister Beatrice

Had Owen survived the war he most certainly would have been involved in the family business possibly expanding the wholesale operation in outport Newfoundland. His death, following on the heels of Herbert's untimely demise from tuberculosis, may account for the senior Steele's decision in 1921 to return to England with his wife and four of his daughters – Mary, Beatrice, Frances, and Ethel – leaving the management of the family concern to his three sons, James, Victor, and Richard, and his daughter Ella. Finding the situation too crowded Richard eventually went off on his own, opening up a wholesale and retail china, glass, and earthenware store opposite the court house on Water Street. Now in direct competition with the family firm, Richard soon found himself in conflict with other family members. In 1925 he married Sybil Hiscock, the daughter of one of the local merchants, a union that proved beneficial to his own enterprise. Eventually Richard reconciled with his two brothers.

The Great Depression proved difficult for small firms like S.O. Steele. This was especially true for the Steeles as the firm was the only source of income for James and Victor and their small families and for Samuel in England, at least until he died in 1936. (Sarah had died in 1925.) After her father's death Ella joined her sisters in England. There she worked as secretary for several firms and during the war years was active in women's groups. Like her father she was a perfectionist but also possessed considerable business acumen. After the war

Inside the shop. Left to right, Miss J. Smith, Miss H. Penny, James R. Steele, Victor J. Steele

she worked tirelessly to create a typescript version of her brother Owen's diary, incorporating letters sent to the family from overseas. Unfortunately, with her death the original letters were lost forever.

None of the daughters married and so the English branch of the family left no heirs. Mary, or Molly as she was called, was a homemaker. The eldest of the girls, she was responsible for attending to the affairs of the family and taking care of Samuel up to the time of his death. She died a year later. Ethel became a nurse and worked in that profession all her life. Beatrice was a homemaker, a most pleasant, generous, and competent lady who worked in her community and church. She was active in Girl Guides and provided myself and my brother, Sub/Lt. Owen S. Steele RN, several holidays when we had brief furloughs during the Second World War. (I served overseas with the 59th Artillery during the final phases of the war in North West Europe, returning to St John's in July 1945.) Frances was best known for her role as assistant manager to Mrs Read, who ran the Newfoundland Caribou Club in London during the last war.[7] The Caribou Club afforded a homelike atmosphere for Newfoundland servicemen like myself during leave periods. I have many happy memories of my time with English relatives during the war years. It was one of the few links my generation had with that branch of the family, apart from correspondence.

James and Victor ran the family business until the 1970s. James died in March 1970. Victor continued to take care of the firm's accounts until he retired in 1976 because of ill health. It was then that the running of the family business passed to the third generation, and I was left with the sole responsibility for the firm's operations.

By the 1970s the business took on a new focus. The wholesale trade with the outports faced fierce competition from local enterprises. It was then that I decided to shift into the retail trade, concentrating on quality English china and dinnerware. With my wife, Francis, and a small staff we managed to carve out a profitable business in this trade. By the late 1980s, however, we were forced to obtain supplies from mainland Canadian wholesalers rather than direct from the Staffordshire potteries. Rising costs, expanding competition from national stores at the retail level, and reduced sales now spelled doom to small businesses like ours. We reluctantly decided to sell out in 1989. Many of our loyal customers expressed their regret at this decision but we were able to pay all our creditors fully and close out successfully. By then the firm had been in existence in one form or another for 141 years – from 1848 to 1989.

100 Water Street still remains, modified somewhat to accommodate a new enterprise but structurally the same building where three generations of Steeles lived out their lives. To what extent we were shaped by my grandfather's enterprise is hard to say but certainly the firm was an important ingredient in all of our lives.

Abbreviations

NCO	Non-commissioned Officers
NI	Not identified
OC	Officer Commanding
PANL	Provincial Archives of Newfoundland and Labrador
PC	Postcard
PMLO	Principal Military Landing Officer
QM	Quartermaster
RAMC	Royal Army Medical Corps
RE	Royal Engineers
RFA	Royal Regiment of Field Artillery
RGA	Royal Garrison Artillery
RFR	Royal Fusiliers (City of London)
RNR	Royal Naval Reserve
RSM	Regimental Sergeant Major
SM	Sergeant Major
SWB	South Wales Borderers
str	Steamer
WPA	Women's Patriotic Association

LIEUTENANT OWEN WILLIAM STEELE OF
THE NEWFOUNDLAND REGIMENT

The Steele Diary and Letters

Historian Will Durant calculated that only twenty-nine years of human history have not been marked by war. British military correspondent Colonel Repington, whose life's work was concerned with reporting on war and military affairs, concluded pessimistically that the history of mankind was the history of war.[1] The frequency of armed conflict in the human experience makes the study of war relevant as an academic discipline and also accounts in large measure for its popularity as a subject, especially at the present time. But this was not always the case. At one time military history was confined to the military academy, where aspiring young officers used it as a field of experience from which to draw important lessons on leadership and the mechanics of command. Those writing in the field served the intended audience by describing campaigns in detail and focusing on the larger issues of command and control. Certainly the big picture is still central to an understanding of the character and complexity of war. But it is not the whole story. John Keegan's seminal work *The Face of Battle* was a reminder of the importance of the soldier's experience to a fuller understanding of the nature of war. Others have also recognized this, notably a generation of historians mining the rich resources of the Great War period.

Wars have a way of generating a body of experiential literature – letters, diaries and memoirs – produced by the more literate, often junior officers, driven by a need to recount or remember an event of significance in their own lives. Little of this literature finds its way into publication. Most remains forgotten, either in private hands or in archival collections. These accounts provide a different perspective on wartime experience – the view from the ground – and can both illuminate and fill in the details of the larger picture. Many have served that purpose already, enlarging our understanding of the nature of war and its impact on the lives of individuals. The American Civil War, for instance, generated an enormous amount of material because its citizen-armies were recruited from all sectors of society, including the more literate. The same occurred during the Great War. That latter conflict also inspired some of the greatest war poets and literary figures of this century, notably Siegfried Sassoon, Wilfred Owen, and Robert Graves. These men became literary spokesmen for the generation that fought in the conflict, giving expression to feelings that most could not articulate and leaving behind an image of war that has become firmly lodged in our consciousness. Paul Fussell's *The Great War and Modern Memory* has shown how our perceptions of war have been firmly rooted in the literary heritage of that conflict.

Not all who wrote about the Great War were the calibre of Sassoon, Owen, and Graves. Hundreds of soldiers wrote personal narrative accounts of their war experience – diaries and memoirs. They are what Samuel Hynes in *The Soldiers Tale: Bearing Witness to Modern War* had termed the "distinct" and "individual" voices that recount the "human" tale of war. He believes that "If we would understand mankind's most violent episodes, we must understand them humanly, in the lives of individuals." Doing so gives modern war a human face.[2] Other historians, such as Martin Middlebrook, Denis Winter, Eric Leed, and Lyn Macdonald, have also taken up that challenge.

Hynes' perceptions are helpful in assessing the place of Owen Steele's diary and letters in the context of soldier's accounts of the Great War. He sees individual accounts recorded in letters or diaries as serving a two-fold purpose – the need to bear personal witness to an event of singular importance in an individual's life and the need to remember for future reference. Diaries and letters have an immediacy that memoirs do not. As he points out, the latter tend to be "reflective, selective, more self-consciously constructed" than the former. As well, time has a habit of dulling memories and softening images. The diary records both the individual's experience and the transformations of character and perspective that occurs over time. The writing in diaries and letters is in a class of its own, with only broad similarities to other forms of literature such as autobiography or traveler's accounts.[3] Owen Steele's diary is the best known of the small number of Great War accounts and correspondence produced by Newfoundlanders in that conflict.[4] Other wartime writing, most notably Mayo Lind's published letters, Richard Cramm's *The First Five Hundred*, and John Gallishaw's *Trenching at Gallipoli*,[5] either cover the same ground or portions of it. But post-war memoirs such as Gallishaw's and those published in *The Veteran* magazine lack the immediacy of Steele's and Lind's accounts.

Steele's and Lind's are the best first-hand accounts we have of the Newfoundland war experience in the first two years of the war. But as Lind and Steele died within days of one another their accounts cover only a portion of the total war experience. It was left to Lt. Colonel G.W. Nicholson to complete the story in *The Fighting Newfoundlander*, the first of two volumes covering the province's military history. Nicholson's history, published in 1964, remains the best work on the Royal Newfoundland Regiment and the province's role in the Great War. It was based on first-hand accounts, published works, and contemporary sources, including the Steele diary and letters. Nicholson himself remarked that Steele's diary "was a valuable contribution to the documentary record of the regiment."[6] Patricia O'Brien's unpublished MA thesis on the Newfoundland Patriotic Association represents the only study to date on the organization of the colony's war effort.[7]

Mayo Lind's account, written to the St. John's *Daily News* as letters from the front, was eventually published as a single collection in the 1920's.[8] Lind's account of the regiment's progress to its destruction on the battlefield at Beaumont Hamel is more carefully crafted than Steele's because it was intended for public consumption. Steele's diary was a personal narrative, intended to provide a record of his wartime experience – to be a reminder, when it was all over, of what war was like and how it felt. As Hynes has pointed out, this was typical of most of the diarists of the Great War.[9] Letters, on the other hand, were intended for the family and, like Lind's correspondence, were written to be informative and entertaining while minimizing the more gruesome side of the war experience. Loved ones were not to be left worried about those engaged in a far-off war.

Steele's account begins with the embarkation of the First 500 on board the *Florizel*. Screening of prospective recruits was so strict that only the best physical specimens were recruited and most came from the capital city. As the population of St. John's was then fairly small, about 40,000, Steele probably had many friends and acquaintances among those selected, either through his connections with the church brigades that supplied many of the first recruits, his sporting interests, or his links with the St. John's commercial establishment. The frequent references to members of the regiment in his letters home suggest a wide circle of friends and acquaintances well known to his family.

Newfoundland Highlanders non-commissioned officers. Corporal Owen
Steele is in the second row, fourth from left

In the beginning the various church brigades, the only para-military
organizations in the colony at the time, provided the largest source of
volunteers for the regiment. The church brigades had their origins in the
popularization of military forms and culture in the pre-war years and
followed the British example, where similar movements had existed
from the end of the nineteenth century.[10] The Church Lad's Brigade,
connected with the then Church of England and the first to form in
November 1892 and also the largest, still exists today. The Catholic
Cadet Corps, the Methodist Guards, and the Newfoundland Highlanders (Congregationalist) followed in succession. The Legion of Frontiersmen, which began in St. Anthony on the Northern Peninsula in
1911, was unlike the other brigades in being non-sectarian, recruiting
broadly from the adult male population, and concentrating its efforts
more on drilling and marksmanship. In its ranks were telegraphists,
schoolteachers, farmers, and fishermen.[11] The Newfoundland Highlanders, to which Owen and his brother Jim belonged, were an important source of volunteers. They had been formed in 1907 and had a reputation for tough summer training that included route marches.[12] Owen
and Jim excelled at the latter. In fact, both were members of the St.
Andrew's Sports Club before the war and won several long distance
walking competitions, popular local events reported widely in the press
that brought them into contact with the city's sports community.

This method of recruiting meant that in the beginning the regiment was closely identified with St. John's, an identification that continued for some time. The officers in these various church brigades, commanded by prominent business and professional men and officered by their sons, also supplied many of the regiment's first commissioned ranks.[13] At first commissioning followed strict quotas based on religious affiliation as well as the individual's social status and political connection. Leadership skills were not a strong factor in these decisions. As Patricia O'Brien points out, "Generally speaking, the Regiment was administered and officered by the St. John's upper and middle class for over three years."[14] Most appointments came from the St. John's Protestant establishment, much to the chagrin of the Roman Catholic community and the outports. This was alleviated somewhat in the second year of the war when the authority for recommending such appointments was transferred to the regimental commanding officer. As the war progressed, battlefield conditions effectively eliminated the social and religious distinctions that had been so important at home.[15]

Owen Steele was not among those first commissioned with the regiment, even though his father petitioned on his behalf. On the surface his credentials looked impressive: he was connected with the mercantile establishment through his family and was well known in the community through his various associations. He sang in the choir in the Queen's Road Congregational Church and served on the Board of Managers with responsibilities for church finances. Although the congregation was fairly small at the time, its members included a number of prominent city merchants.[16] Steele's family was also connected through marriage with the firm of J.H. Martin and Company, a successful hardware firm in St. John's. While S.O. Steele and Company was not in the same league as the great commercial firms that dominated the colony's economic life, they were part of the small but thriving business class of the city. As the firm's travelling salesman, Owen had business contacts with the merchant community outside St. John's, especially on the Burin Peninsula. He had also served with the Newfoundland Highlanders although as a Sergeant and not as an officer. However close his connections with the local establishment might have been, they were clearly not strong enough to warrant consideration by those responsible for the management of the regiment in the early months of the war. As a result Steele began his military service on 13 September 1914 as Private Owen Steele, Regimental No. 326, although within a month he was promoted to colour sergeant. His commissioning came less than a year later when appointments were based on merit rather than on connections at home.[17]

Once Steele was commissioned his contact with other ranks was governed by the protocols of military service. His duties and responsibilities placed him in command of others, thereby altering earlier relationships. His social life revolved increasingly around the officer's mess. Even on leave, whether extended or restricted, he usually associated with brother officers. Occasionally he and Jim travelled together when they visited English relatives. As an officer, Owen's view of the regiment's activities represented a slightly different perspective than the soldier's. This can be seen if Steele and Lind's accounts of similar events are compared. Nevertheless, while the impressions of events might differ, the experience is still a shared one.

Steele's account of the war is limited in time to the first two years of the war. Much of its content relates to the months of training that preceded the regiment's first engagement at Gallipoli when the regiment's involvement was peripheral to the main theatre of operations, in which other British, French, and ANZAC forces suffered terrible losses. Gallipoli, however marked an important transition for the regiment and for Steele. The regiment was now part of a regular British division, the 29th, where its performance was measured against the grueling standards of the professional soldier. Gallipoli was the regiment's baptism of fire. It was at Suvla Bay that it suffered its first casualties. It was also at Gallipoli that it earned the reputation for steadiness and reliability that followed it throughout most of the war. Steele shared that experience and played his own role in the highly successful withdrawal from the Peninsula. Gallipoli demonstrated the harshness of war, the personal hardship and deprivation, the ever-present reality of death and suffering. The regiment was transformed by that experience. So too, was Steele. Gone was the buoyant optimism and sense of adventure that marked the first year of the war. The diary entries after that take on a different tone.

After Gallipoli Steele's entries represent the more serious side of the war experience. His accounts become more graphic and more expressive. This is especially the case when the regiment moves to France and begins planning for the Battle of the Somme. Unfortunately, Steele's account ends on the eve of the battle. Although scheduled to go over the top that fateful morning, at the last moment he was held back as part of the ten percent reserve. We will never know how he reacted to the destruction of his regiment on that day as he apparently never had time to make another entry or write another letter before he was killed less than a week later by an exploding shell fragment in an area behind the front lines. The last thing in the diary is an account Steele inserted of how many of the men of D Company returned unwounded from the battle on 1 July. Steele's account of the Great War is therefore not complete. But as Hynes rightly observes: "so much of what actually happens in war is unrecorded."[18] The surviving accounts give us occasional glimpses of man's war experiences. Steele's diary and letters provide just this.They have an immediacy and directness that no historical account can possibly provide.

There is no indication that Steele ever intended to publish his diary and letters. The diary was his personal record of his wartime experience. Like so many who rushed to the colours in 1914, Steele was captivated by a sense of adventure, eager to be part of events of great importance, to intersect with what Hynes calls a "world of great doings."[19] But, as Steele's diary reveals, this encounter with great events was slow in coming. Gallipoli should have served that purpose but the Newfoundland Regiment's role in the operations was minimal and the hasty withdrawal simply accentuated the overall sense of failure attached to the enterprise. It did introduce Steele to the harsh realities of war and provided a brief moment of well-deserved fame for his part in the carefully planned and executed withdrawal from Suvla Bay. France held greater promise. The planning for the Somme offensive buoyed up Steele's spirits but also intensified his sense of apprehension. The outcome could bring about the much-hoped-for end of the war but at what cost? Lives would be lost. Would his be one of them?

Our encounter with Owen Steele ends on the eve of Beaumont Hamel. What we see through the diary and letters, in a sense, represents the complete man – all that he has ever been and all that he will ever be. Steele was typical of the generation that marched off to fight for God, King, and Empire in 1914. He shared their youth, their ideals, their optimism, their sense of adventure and commitment to a patriotic cause. Nothing in civilian life could have prepared Steele for what war would be like but, like his comrades, he quickly adapted to the rigors of military life and by 1916 was in every respect a seasoned soldier. He was twenty-seven when he entered the service, his character and personal traits already well formed. The man we come to know through the diary was probably very close to what he had been as a civilian – intelligent, hard working, a good team player with an outgoing and buoyant personality. He also comes across as competent and efficient, traits no doubt acquired while working in the family business and the kind of qualities that were quickly noticed by those in authority. Hence the increasing complexity of duties assigned and the rapid promotion to Second Lieutenant once the regiment was up to strength. The diary also reveals something about Steele's sensitivity – his caring attitude towards family and friends, his efforts to allay the fears and concerns of those at home, his shock at the horrors of war and the carnage it left in its wake. Publicly, in his letters home, he remained confident and optimistic; privately, in the pages of his diary, he expressed a growing weariness with war, a strong desire to return home, and an underlying hope that he might avoid a soldier's death.

Nicholson wrote of Steele: "A soldier of sterling qualities, during his progress in the Newfoundland Regiment from N.C.O. to acting Company Commander, his foremost care had ever been for the welfare of his men. He took keen interest in all that was going on about him, and the daily diary which he kept so conscientiously and efficiently from the day of the First Five Hundred's embarkation until the eve of Beaumont Hamel was to make a valuable contribution to the documentary record of the regiment."[20]

NEWFOUNDLAND AND THE GREAT WAR

Much of the lore about Newfoundland and the Great War focuses on the activities of the Royal Newfoundland Regiment, in particular the action at Beaumont Hamel on 1 July 1916 when the regiment was virtually wiped out in the opening phase of the Somme offensive. Over the course of the war some 12,000 Newfoundlanders joined the fighting services – navy, army, air force, merchant marine – and nearly a third of these made the supreme sacrifice.[21] On the home front, thousands of Newfoundlanders worked directly and indirectly for the war effort, many in voluntary associations, but their efforts have been largely forgotten. Instead, the focus has always been on the regiment and its distinguished record of service as part of the fabled 29th Division of the British Army. The regiment served in two major theatres of the war – Gallipoli and the Western Front – and its battle honours include Albert (Beaumont Hamel), 1916; Le Transloy, Arras 1917; Ypres, 1917, 1918; Langemarck, 1917; Poelcappelle, Cambrai, 1917; Bailleul, Courtrai, Gallipoli, 1915–16. Members of the regiment won scores of decorations and produced the youngest VC in the Empire. In February 1918, as a singular honour, conferred on no other British unit during this war, the Newfoundland Regiment was awarded the prefix "Royal" by King George V in recognition of its "magnificent bravery and resolute determination" on 28 September 1917 in the Ypres and Cambrai battles.[22] Although the Regiment's losses at Monchy-le-Preux (April 1917) were almost as devastating as those at Beaumont Hamel, it is Beaumont Hamel – Newfoundland's first encounter with mass death during wartime – that has been etched deeply into the consciousness of Newfoundland and has come to form part of the mythology of a land and its people, a symbol of its proud history and of its supreme sacrifice for King and Country.

In 1914 Newfoundland was not a part of Canada. During World War One the island of Newfoundland and the mainland territory of Labrador together comprised what was then called Newfoundland. A self-governing colony enjoying the same degree of independence over her internal affairs as the neighbouring Dominion of Canada, Newfoundland had participated in the initial round of meetings that produced the Canadian confederation in 1867 but rejected the terms of union at that time and in the years before the First World War. Newfoundlanders saw few advantages to be gained from union with Canada[23] and generally perceived themselves as different from Canadians. For one thing they had a better sense of identity than their Americanized cousins on the "Mainland" and possessed a distinctive culture that represented an amalgam of Celtic and English West Country traditions. They were British and inhabitants of Britain's oldest colony.

Newfoundlanders were an island people, mainly fisherfolk, whose lives and livelihood were shaped by the icy tempestuous waters of the North Atlantic. For centuries those waters were seen as a nursery for sailors in Britain's Royal Navy.[24] But they were also the graveyard of many of the colony's fishermen. The spring seal fishery was particularly notorious. In March 1914 the SS *Southern Cross* disappeared with a loss of 173 hands. Less than a month later, seventy-eight of the crew of the SS *Newfoundland* were caught on the ice and perished in a blinding snow storm.[25]

Sharing a common Anglo-Irish ancestry, Newfoundlanders had a strong sense of community, whether "townies," as the residents of the colony's capital and commercial center St. John's were called, or "baymen" living in the small communities along the colony's rugged coastline. Traditional antagonism between the two reflected the dominance of city over outport that had characterized the island's economic existence over the previous century.[26]

Newfoundlanders were a deeply religious people with an abiding faith that carried them through times of great difficulty. St. John's boasted some of the finest examples of Romanesque and Gothic architecture in British North America. In the outport communities magnificent wooden structures, some the size of small cathedrals, dominated the shoreline. Religion and sectarian interests exercised a powerful influence over the colony, shaping its political life, controlling public education, and even functioning as a form of local government.[27] Although an unmilitary colony, Newfoundland had a proud tradition of service to the empire, supplying the fleet with skilled seamen and sending recruits to join Britain's armies in their wars against the Americans in 1776 and 1812 and against the Boers in 1899.[28] Efforts to raise volunteer corps in the 1860s had been largely unsuccessful[29] and when Britain withdrew its imperial garrison in 1870 the island was left defenceless except for guarantees offered by the Royal Navy and the presence of a small local constabulary. Nevertheless, there was considerable local interest in military affairs and the colony continued to support a small branch of the Royal Naval Reserve (approximately 600 men), based in St. John's but recruited mainly from the outports, that offered training in seamanship and an opportunity to share in the proud traditions of the Royal Navy on board the unit's training ship HMS *Calypso*.[30] The city brigades also provided a form of military training.

On the eve of the war Newfoundland had a population of a quarter of a million, 45,000 of whom lived in St. John's, the colony's chief commercial and financial center. The rest were scattered in some 1,300 small settlements that dotted the 6,000 miles of rugged coastline of the island and Labrador.[31] The colony's annual revenues were modest, just over $3.6 million annually; its funded public debt amounted to $30.5 million. Although it was rich in natural resources, minerals and timber in particular, it was the fishery that provided the underpinning of the island's economy and the principal livelihood of its inhabitants. Lacking the sophisticated technology that in our time has depleted the ocean's rich harvest, the Newfoundland fishery of the beginning of this century was a labour-intensive and family-run enterprise.[32] After a very brief downturn on the outbreak of hostilities the island's economy flourished during the war years as inflated prices for fish brought about an artificial prosperity.[33] But the First World War was expensive for Newfoundland, costing the colony on average $100,000 a month to support the Regiment, an expanded Royal Naval Reserve, and a smaller Forestry Corps. Some have blamed the financial collapse of the early 1930s and the subsequent abandonment of Dominion status in favour of a British-appointed commission of government on the long-term effects of wartime expenditure.[34]

When Britain went to war on 4 August 1914 her far-flung empire, including Newfoundland and the neighbouring Dominion of Canada, were automatically drawn into the conflict. "The British Empire is at War and All Europe Ablaze" proclaimed the headline of the *Daily News* in a moment of journalistic excess.[35] The Newfoundland response to the outbreak of war was especially enthusiastic in St. John's, where the connections with Britain were strongest. In the innocence and euphoria of the moment all that mattered was that Britain's oldest colony should respond quickly and favourably to the mother country's call for assistance. While the Royal Naval Reserve practiced manoeuvres in Cuckold's Cove, Sir Walter Davidson,[36] the colony's British governor, proceeded to organize the colony's military response.

In what many considered a "bizarre abdication of responsible government," the colonial government turned the direction of the war effort over to a group of private citizens under Davidson's leadership.[37] For the next three years, until it was replaced by a Department of Militia, the Newfoundland Patriotic Association took on the onerous and increasingly difficult task of raising, equipping, transporting, and caring for the land contingents dispatched from Newfoundland for service overseas.[38] A unique arrangement in the annals of imperial history, it nevertheless worked surprisingly well in a colony with no recent experience in military affairs.[39]

Initially, the Association proposed to raise 500 men for service overseas and to increase the size of the Naval Reserve to 1,000. The response to the proclamation of 22 August calling for volunteers was overwhelming. Within days 335 men had signed up, two-thirds coming from the city brigades. By the end of the first week, it appeared that the entire 500 might be made up from St. John's. When enlistment tapered off by 26 September nearly 1,000 volunteers had signed up. Half passed the required medical exams and moved to tent lines established at nearby Pleasantville.[40] These were the First Five Hundred, a rag-tag little army of enthusiastic volunteers, from the sons of the city's elite, attired in a variety of military costumes, including blue puttees as no khaki cloth was available from local suppliers – hence their more popular designation "The Blue Puttees."[41] The terms of enlistment were "for the duration of the war, but not exceeding one year."[42] By war's end, over four years later, a total of 6,241 Newfoundland men had served in the Regiment, 4,668 as volunteers (Another 5,747 enlisted in the Royal Naval Reserve, the Forestry Corps, the Canadian Expeditionary Force, and British Forces.)[43]

With less than a month's training under their belts, the First Contingent left St. John's for England on board the sometime sealing ship ss *Florizel* on 4 October, joining a convoy of Canadians off Cape Ray.[44] They had received a "right royal send-off" from the largest gathering of citizens ever assembled in St. John's.[45] The Newfoundlanders were eager to get under way as most expected it would be a short war. Even the early cables received on board were optimistic: "German Cruiser sunk off German Coast – Belgians win great victory over Germans. Everything going favourable to allies." But before long the news became less encouraging and the realization struck home that the war would be long and costly.[46] Beaumont Hamel and the enormous casualties it produced represented a turning point for Newfoundland in the Great War in the same way that the Battle of the Somme transformed the character of the war for the British. Thereafter it became a seemingly endless slog, a constant drain on the resources of a nation and its people. The war forced Newfoundland to set aside normal democratic practice and create a national government while extending the life of the legislature until after the conflict was over. The Newfoundland Patriotic Association, that novel experiment in civilian administration of the war effort, proved too susceptible to promoting individual self-interest over public good and was replaced by a Department of Militia. Conscription, or the threat thereof, reopened old political wounds and sharpened divisions between city and country. At stake was the continued existence of a Newfoundland Regiment in the British line,[47] a matter of great pride to most. The war ended before any conscripts were sent overseas but the issue of compulsory service remained contentious for some time thereafter.[48]

The war left a lasting legacy. Patricia O'Brien has summed it up this way: "The overwhelming financial burden of the war, the capitalized cost of which was expected to reach $35,000,000, the staggering loss of life, the unhealthy expansion in the fishing industry and the fact that, unlike Canada, the colony had not been in a position to broaden its basic productive capacity, all took their toll in the ensuing eras of economic nationalism and world-wide depression."[49] And yet, as she also concedes, the war brought Newfoundland a number of honours, including admission to the Imperial War Cabinet, elevation to Dominion status, and the title "Royal" for its beloved regiment.[50] When the Newfoundland War Memorial was unveiled on 1 July 1924, eight years after that fateful morning at Beaumont Hamel, it was Field Marshal Earl Haig who performed the ceremony. To Newfoundlanders it was recognition, at the highest level, of their contribution to the Empire's war effort. Haig's wartime reputation has dimmed over time but Newfoundland's has not. In the annual ceremonies marking the Battle of Beaumont Hamel in Newfoundland and at the magnificent caribou memorial on the battleground, now a national historic site, the heroic efforts of a small country in that great conflict will be forever remembered. Owen Steele's diary will now form part of the public record of that event, a further reminder that war engages not only countries but also individuals.

CHAPTER ONE

Pleasantville to Salisbury Plain
(3 October 1914 – 12 December 1914)

The diary begins on 3 October 1914 with the embarkation of the first contingent of the Newfoundland Regiment on board ss *Florizel*. The 3,081 ton *Florizel* was the flagship of the New York, Newfoundland and Halifax Steamship company, owned by Bowring Brothers, one of the great merchant houses of St John's.[1]

The departure of the *Florizel* was an occasion for great excitement in St John's as the largest group of citizens ever assembled gathered at the Furness Withy Company's pier to give the regiment a "right royal send-off." Cheering crowds had accompanied the regiment's march from the make-shift training camp at nearby Pleasantville. The weeks of preparation, marked by a daily routine of drill and exercise, had heightened public interest and built up anticipation among the volunteers. It was a rag-tag little army that marched off to war that day, full of confidence and bursting with pride. It was, after all, the first time Newfoundlanders had been called into service since the war with the Americans a full century before – although very few would have been aware of this. Military ardour since then had had brief manifestations – the volunteers of the 1860s and the various paramilitary brigades that sprang up at the turn of the twentieth century. These uniformed brigades, with the exception of the Legion of Frontiersmen, had a connection with local churches and were generally St

John's–based. With their emphasis on parades and drills, these corps built up a respectable following in the years before the war, providing the colony with much needed experience in military affairs. Most of the "First 500," as they came to be known, had some association with these units.[2]

The ship finally left port on the night of 4 October, two months to the day after Britain declared war on Germany. It took ten days to make the crossing. Two days out, Steele was promoted to colour sergeant or company sergeant major. It was a proud moment for the still-green recruit but a token of the favourable impression he had already made. He was a good organizer, a natural skill he seems to have developed in his youth. His experience in the church choir of the Congregational Church made him an obvious choice as program director for a regimental concert organized to break the monotony of the long and wearying trip across the stormy north Atlantic.

Landfall at Plymouth, in the historic west country, the ancestral home of many Newfoundlanders, was a relief for all, especially the many who had been seasick during the crowded and uncomfortable voyage across the North Atlantic. For nearly six days the men remained on board the *Florizel* – only the officers and a few of the NCOs enjoyed a bit of shore leave. When the Regiment finally disembarked and began the eight-hour train trip to their barracks, they were enthusiastically welcomed by everyone they encountered.

Salisbury Plain was the great training ground for British and Empire forces, its vast chalky downs ideal for large-scale manoeuvres but likely to be sodden from too much rain. The monotony of life aboard ship was replaced by the monotony of training – daily parades and route marches designed to harden up and discipline fresh recruits. Formal inspections by visiting dignitaries broke the routine, as did the occasional tactical exercise against imaginary enemies in abandoned farm houses. By late November the earlier dampness and mud gave way to the bitter cold of the exposed plain. The diary reveals the extent to which so much of Steele's life at this point was burdened by a constant sameness: "Nothing unusual to-day" is a typical entry. It was difficult for Steele to find something interesting to say in his letters written from Salisbury Plain, although he certainly made an earnest effort to do so. Lengthy descriptions of the daily routine and of the vast tent city that was now home betray the growing tedium, broken only by nightly entertainment in the mess or by weekend excursions. Homesickness was occasional but at this early point not a serious issue either for Steele or for his comrades.

At Salisbury Plain the Newfoundlanders learned the soldier's trade but also nurtured a developing sense of unit solidarity and national pride[3] as the larger Canadian presence at the camp created many occasions for mistaken identity. Far more serious was the fact that the Newfoundland Regiment was under-strength and faced the possibility of amalgamation with either a Canadian or a British unit, which would have led to a complete loss of its identity. Steele struggles with the issue of identity in the first pages of the diary. The points he makes are small ones but the search for defining characteristics or notable differences shows Steele's preoccupation with this concern and probably represent the feelings of most in the regiment. The war went a long way to developing latent feelings of nationalism in the Newfoundland contingent but always within the wider "British" context.

News that the regiment was to be augmented and to proceed into barracks at Fort George, close to Inverness, Scotland, was a welcome relief as reinforcement meant that the regiment's identity would be preserved. Steele saw the new assignment as indicative of the high standard achieved by the regiment in its training at Salisbury Plain. A visit by Lord Brassey, a former lord of the Admiralty and an English member of the Newfoundland War Contingent Association, confirmed to Steele and others that the regiment has already achieved special recognition from England and the English. Prophetically, Brassey had remarked, as Steele recorded, that the regiment would bring credit to itself and to Newfoundland.

SATURDAY – OCT.3/14
Left Pleasantville at 4 p.m. and called on the Governor.[4] Then after a short march arrived on board S.S. *Florizel*[5] about 5.30 p.m. Left F.W.[Furness Withy] & Coy's wharf about 7 p.m. Anchored in the Harbour.

SUNDAY – OCTOBER 4/14
Still anchored in the stream. Besieged by Boats filled with visitors all day. Left port at 10 p.m.[6]

MONDAY – OCT. 5/14
Sighted the ships of the Canadian Contingent about 10 a.m.[7]
Reached them about noon and our ship was given its allotted place.[8]
We then all commenced our long trip to England. There were about 35 vessels in sight. It was a glorious sight.

Orders were received to have "lights out" at 7 p.m. All port holes were covered and we were nearly all in bed by 7.30 p.m. – I was.

TUESDAY – OCT. 6/14

Got up at 7.45 am, having had 12 hrs sleep. Still quite windy, as when we started, but ship fairly steady. Rather foggy but cleared a bit at dinner time. After dinner had another fine view of the steamers in sight – counted 35.

This evening several promotions were made known to date from Oct. 7 amongst them being colour Sergt, or Coy. [Company] Sergt. Major for self.[9]

WEDNESDAY OCT 7/14

Ship rolling quite a lot this morning and continued so all day, impossible to keep dishes etc on meal tables.

One of our fleet stopped a steamer going in another direction, but after a short while she was allowed to continue passing near us. She was a Norwegian. A lumber str.[steamer] was also met. At night weather still cold and windy and steamer rolling a lot.

THURSDAY – OCT. 8/14

About noon we left 4 strs behind for some unknown reason, regulated speed of fleet was about 9 ½ miles per hr. A battle ship (or cruiser) went back later to escort them. Abt. 6 p.m. the "*Laurentic*", which had spent all day away off to the north flank reconnoitering returned, and wheeled around just behind us and many cheers were exchanged, only 100 yds separating us – her decks were crowded. At night str. still rolling as for past couple days.

FRIDAY – OCT 9TH/14

The steamers left behind yesterday were again with us today. Abt. 11 o'c. saw many porpoises playing in water. Passed 3 or 4 dead horses (have passed some every day). [10] Saw a large sailing vessel in the distance abt noon. Shortly after tea received following war news from Flagship: "German Cruiser sunk off G.[German] coast. – Belgians win great victory over Germans -Everything going favourable to allies." Greeted with cheers.

Spent tonight preparing for concert.[11]

Our course, which has been changed several times, was at 8 p.m. changed completely to right angles. Owing to change of direction str. now going much steadier.

SATURDAY – OCT. 10TH/14

At 3.20 am, the bugle sounded the fire alarm, and we all had to turn out on deck dressed. It was a test alarm.

About noon we met a large str. towing an oil-tank str with five mast[s] – she was using oil as fuel, the smoke coming out of her foremost mast. A large Battle Cruiser was on the horizon to our port bow all day. At 6 p.m. *Laurentic* again returned after 2 hrs cruising.

Spent tonight preparing program & practising various items thereon.

SUNDAY – OCT. 11TH/14

Catholics held service at 10.30 a.m. under charge of Captain Carty.[12]

Protestants held service at 11 a.m. – conducted by Revd Stanlake [Stenlake],[13] assisted by Dr. Wakefield[14] the former delivering a sermon. Service held on upper deck for which the piano had been moved up. After tea a two hours singing service was held – a large number attended.

Today the weather was glorious – ship still rolling as wind still continues.

The *Laurentic* on her return this evening passed close behind us and cheers and songs were exchanged.

Spent afternoon preparing programs &c for concert.

MONDAY – OCT 12/14

Today we had the pleasure of seeing one of England's latest Battle Cruisers the *"Princess Royal"*[15] which has for the past couple days been on the horizon guarding our left flank. She wheeled around quite close behind us. We gave her three cheers and her band responded with the new Canadian Anthem evidently putting us down for Canadians. We responded with "Britannia Rules The Waves." Several cheers were again exchanged.

Shortly afterwards about 6 p.m., the *Laurentic* returned from her daily scouting trip, when cheers were again exchanged.

At 7.30 p.m. a regimental concert was held, having been promoted by the writer. There were some 25 numbers and all voted same an immense success.[16]

TUESDAY OCT. 13/14

Very windy indeed today; and all parades cancelled on a/c [account] ship rolling so much. Saw a couple strs today.

Nothing else of importance today.

Recd. discouraging war news by wireless.[17]

WEDNESDAY OCT 14/14

Sighted land (presumably Land's End) at 10 a.m. – first land seen since leaving Narrows 10 p.m. Sunday Oct. 4th. Saw a barquentine

REGIMENTAL CONCERT

Under the Patronage of Captain [Conn] Alexander
Officer Commanding H.M. Transport *"Florizel"*
- oOo -

1.	Vocal Solo	"Anchored"	Lce. Cpl. Williams
2.	" "	"Just let me like a Soldier fall"	Pte. Gordon Greene
3.	Instrumental	(Banjo	Drum Major Miller
	Quartette	(Accordeon	Col. Sgt. S. Ebsary
		(Mandolin	Corporal Mercer
		(Clarionet	Pte W.L. Norris
4.	Vocal Solo	"His Day's Work Was Done."	Private E. Pike
5.	" "	"Beef Stew & Carrots"	Bugler W.P. Kenneth
6.	Double		
	Quartette	"Old Black Joe"	Florizel Glee Singers
7.	Vocal Solo	"Asleep in the Deep"	Pte. H. Ebsary
8.	" "	"Norwegian Sea Song"	Pte. Lewis
9.	" Duet	"Life's Dream is o'er"	Ptes. Greene & Spencer
10.	Instrumental		
	Solo	"Clarionet"	Pte. W.L. Norris
11.	Vocal Solo	"My Old Kentucky Home"	Lt. H. Goodridge
		and chorus	Florizel Glee Singers
12.	" "	"Thoria"	Pte. J. Spooner
13.	Double	"My Bonnie Lies Over	
	Quartette	the Ocean"	Florizel Glee Singers
14.	Recitation		Pte. G. Goodyear
15.	Vocal Solo	"Drink to me only with	
		thine Eyes"	Lce. Cpl G. Byrne
16.	Instrumental	(Banjo	Drum Major Miller
	Trio	(Mandolin	Col. Sgt. S. Ebsary
		(Mandolin	Corporal Mercer
17.	Vocal Solo	"Tipperary"	Lce. Cpl. C. Clift
18.	Instr. "	"Accordeon"	Col. Sgt. S. Ebsary
19.	Vocal "	"The Perfect Day"	Pte. A.L. Summers
20.	Recitation	"On the Dawson Trail"	Lce. Cpl. R. Bartlett
21.	Vocal Solo	"I'm absolutely full"	Lce. Cpl. S. Ferguson
22.	Instr. "	Clarionet	Pte. W.L. Morris
23.	Vocal "	"The Maple Leaf Forever"	Lce. Cpl. J. Robinson
24.	Vocal "	"The Marseillaise"	Capt. A.E. Bernard
25.	Ensemble	"Auld Lang Syne"	Ensemble

GOD SAVE THE KING

1st *Tenor*	2nd *Tenor*	1st *Bass*	2nd *Bass*
Lce. Cpl C. Clift	Pte. J. Spooner	W.H. Grant	Pte. H. Ebsary
Pte. G. Greene	Pte. R. Patterson	Lce. Cpl Williams	Col. Sgt. O. Steele

which crossed our bows & passed within a few feet of us. She was a large four masted French one & dipped her Flag in our honour. Saw several steamers today. Reached vicinity of Plymouth abt [about] 5 p.m., & found after going part of the way in, that there was no room for us at present, so we went out again and remained steaming about & at midnight were still moving around.

Got discouraging war news again today.[18] Very busy day for me as Col. Sergt. Wrote letter home late tonight.

EXTRACTS FROM A LETTER DATED 14 OCTOBER 1914, WRITTEN ON BOARD HM TROOPSHIP *FLORIZEL*

This is to be our last night on the water, I believe, for we are now lying off Plymouth, which Port we approached about 5 p.m. this afternoon. We were ten days on the journey across the Atlantic. We are not to mention anything regarding the movements of the Convoy.

On the 12th inst. I was inoculated against Typhoid, when 500,000,000 typhoid germs were put into my body, and ten days later, (the 22nd), 1,000,000,000 more are to be put in. I believe about two-thirds of the whole Regiment have been inoculated during the trip, nearly all of whom have been very much troubled with sore arms, biliousness, sick headaches or something. One morning, in our Company some twenty-three or twenty-four men were absent suffering from effects of inoculation or seasickness. The ill effects last for forty-eight hours, but am pleased to say, that beyond my arm being slightly tender (in comparison to others), I have suffered no ill effects whatever. They say this is proof that one is in a healthy condition, the health germs, or disease resisting germs in one's body having the predominance of power. I am glad if this be the case, for in time of war, many times as many men die from disease as from bullets. In the Boer War seventeen died from disease to one from bullets. This, of course, is an exceptional case, but the rate is usually eight or ten to one; so I stand a very good chance of keeping clear of disease, even if I do not escape the German bullets.

By the way, quite a large number of our men suffered from seasickness, including some half dozen or more of the officers.

I will endeavour to cable upon arrival.

THURSDAY – OCT 15TH/14

Came into Plymouth again at 10 a.m. and anchored. At 4 p.m., weighed anchor and continued up to Devonport & dropped anchor at 4.30 p.m. On the way up the Sound saw the training ships *Impregnable* ... of the Nelson type. Received a one gun salute from the *Impregnable* – all were crowded with cadets. Beautiful scenery.

Most of the transports are either at the Piers or anchored in harbour around us.

Capt. Franklin[19] came aboard at 6 p.m. – He took my letters (4) to post them & told me there was no need to telegraph home as he was telegraphing the Govt. We were all anchored in pairs – two steamers to a buoy. Heard tonight that we were likely to remain here for five days.

On the way from Plymouth we were greeted with cheers every-where – mostly from ladies.

FRIDAY OCT. 16/14

Still lying in Devonport Hr. No one allowed on shore, though some have gone on shore from other steamers. During the day several large Tugs, pleasure boats & water boats cruised around the various steamers waving handkerchiefs etc., and shouting "are we down-hearted?" to which the answer sent back was always "No." Cheers were continuous. No one of those who do go ashore are allowed to pass thro' the docks to go aboard steamers after 6 p.m. (This is to guard against spies). Our buoy-partner the *Corinthian* which seems very much larger than she used to be, is alongside of us.

SATURDAY – OCT. 17TH/14

Still in Devonport Hr. This morning some of the officers went ashore for the day but do not think we shall get a chance. The steamers are going to the wharf one by one, but our turn has not yet come.

Today has been a repetition of yesterday. We had no 2.30 parade owing (we believe) to the fact that Sat. afternoon is half holiday in Eng[land].

[*Note: The following text is only in the typescript copy of the diary and letters*]

This evening we heard that Canada and Newfoundland heard five days after we left St. John's, that a certain number of ships of the Convoy had been sunk by German Battleships which were encoun-tered in the Atlantic, and that St. John's was in mourning as the *Florizel* was stated to have been amongst those gone down, and they knew nothing different until Capt. Franklin telegraphed our arrival on the 15th inst. I am pleased to say that we had no trouble of that sort during the trip, in fact, our passage was almost *too uneventful*. I consider the fact of our getting such a large convoy of steamers across the Atlantic in safety to be nothing less than a "Naval Victory".

SUNDAY, OCT. 18TH/14

The Catholics had service on deck at 10.30, under direction of Capt. Carty.

The Protestants held their service on deck at 11 o'c. under Dr. Wakefield assisted by Revd Stenlake. I was asked to have the Florizel Glee singers in attendance to lead in the singing & as a result we had a fine choir. At 2 p.m. we commenced sending the Regiment ashore to go on a route march & spent about 2 hours on a march through various portions Devonport. It is rather a quaint old place, but still we would all like to have a little leave but cannot get 5 mins off – too bad.

It was about 7 o'c. by the time all were on board again. After tea we had some Hymn-singing.

MONDAY – OCT. 19TH/14

Nothing occurred during the morning, but after dinner the Colour Sergts & one Sergt from each Coy. were given leave until 6 p.m. I got away at 3.00.

Went over to Plymouth & bought wristlet watch. Walked around a little & then it was time to go on board again. Did not have time to see anything still was pleased to get the little [leave], since it was the first leave I had had since Oct. 2nd.

TUESDAY – OCT 20TH/14

Got up at 6 a.m. & had breakfast at 8 a.m. & immediately made preparations for disembarking. At 12.30 left *Florizel* and at 1 p.m. started on our march to Plymouth. Arrived Plymouth 3 p.m. and think we just missed our train. So continued on a couple hundred yards to the grounds of the 5th Devonshire Reg. to await next train at 7.15. – Reg. was given permission to roam grounds & Sergts & C. Sergts. were invited to the S. Mess & treated to Billiards &c. – Being on Colour Party had to remain by them but got an hour off later. We left the grds. [grounds] at 6.30 after the Sergts. & C-Sergts had signed a roll for the Devon. Reg. Left Plymouth by train at 7.15. A Mr. Pinsent introduced himself to the Colour Party as Nflder. Also H.W. LeM's [LeMessurier[20]] daughter & her husband. We were all given lots of apples, papers &c. by enthusiastic people & similarly treated at various stations on the way. Arrived at Patney shortly after midnight & in a short while commenced our 7 or 8 mile tramp to Pond Farm Camp, where we arrived shortly after 3 a.m. Wednesday.[21]

WEDNESDAY OCT. 21/14

Arrived at Salisbury Plain Camp shortly after 3 a.m. After a mug up we turned in about 4.30 & got up at 8 a.m.[22] Spent the day straightening off. Recd. letter Aunt Nell.[23]

THURSDAY – OCT. 22/14
Spent today fixing tents &c, also putting floors in tents, & fixing off the Coys.[24]
(Wrote Aunt Laura.[25])

FRIDAY – OCT 23/14
Spent most of today at Pay List. Our 5 Coys were today put into 2 Coys each divided into platoons as per new Drill Book.
The funeral of a Sergeant took place from the Camp next us today. He dropped dead after playing football & the cortege was led out of their camp & past ours by their Band which played the "Dead March."
(Wrote Aunt Nellie.)

SATURDAY – OCT 24/14
Very rainy day – Lord Roberts[26] reviewed us about 12.30, but owing to the rain no one saw him, seeing only the car in which he rode.
Being Saturday afternoon we had the evening off.

SUNDAY – OCT 25TH/14
Had service on the grounds. Three or four other regiments were present just when finished ... arrived & gave us a short address.

MONDAY, OCT 26TH/14
Nothing unusual occurred today, except that Bert Smith[27] came over and called at my tent & stayed for half an hour.

TUESDAY – OCT 27/14
Today I met Ed. [C.E.A.] Jeffery[28] who came over to see the Nflders.

WEDNESDAY – OCT 28TH/14
Today saw Ed. [C.E.A.] Jeffery who came to our Camp.

THURSDAY – OCT 29/14
We went for a route march today of some 15 miles in full equipment.

FRIDAY – OCT 30TH/14
Nothing out of the ordinary today.

SATURDAY – OCT 31ST/14
Half holiday again, so I & 3 other Sgts. arranged for leave but did

not get off until 5 o'c. – Went to Market Lavington & got back again at 9.30 p.m.

SUNDAY – NOVR. 1ST/14
Service at 11 a.m. but as no minister turned up where the Presbyterians met with whom I went we were dismissed with a prayer from Dr. Wakefield. We were glad as it commenced to pour rains.
Wrote ...
Geo. Cowan [NI] called at Camp today.

MONDAY – NOVR. 2ND/14
This morning we marched some 3 or 4 miles to West Down South where a rehearsal was held in preparation for the King's visit on Wednesday. It was pouring rain all the 4 hours we were away & all got soaked. Returned about 1.30 & were freed for the day.

TUESDAY – NOVR 3RD/14
We were given a whole holiday today – being the Queen's Birthday, so I got leave to go to Devizes to a Dentist – Grant Patterson[29] came with me for same purpose. Had to walk 5 miles to Lavington then took train to Patney & transferred & reached Devizes at 2 p.m. Went to Dentist to make appointment but was filled up for the day; asked us to call 3.30 p.m. – Went to Hotel for dinner and after stroll went to Dr. Armina [the dentist]. Had time to put one temporary filling in for me only. Left Devizes by Motor Bus for Mkt. [Market] Lavington at 5 p.m. Had tea in M.L. & left for Camp about 9 o'c.

WEDNESDAY – NOVR 4TH/14
All troops in our vicinity were to be reviewed by King & Queen with Roberts & Kitchener. Our regiment had to walk about 3 to 4 miles. I could not go on account of not having my uniform yet, therefore acted as Bat. Orderly Sergeant and was left in charge of Camp. Nearly 10 others had to remain also for various reasons.
Review took place at 12.30. Regt. returned to Camp 2 p.m.

EXTRACTS FROM LETTER DATED 4 NOVEMBER 1914
We get up at 6 a.m. every morning and have to light our candles to dress by, and have to be on the Parade Ground at 6.15. Wash Parade is at 6.45 a.m., breakfast at 7 o'clock, and then Parade for drill again at 8.30 until 10 o'clock. At 10.30 a.m. until 12 o'clock we have another Parade, then dinner, and Parade again at 2 p.m. to 4.30 p.m. Tea is at 5 p.m., after which we are free until 9 p.m. when the roll is called and we all go to our tents. The bugle for "Lights out" is sounded at 9.45 p.m.

On the way from Plymouth, a black cat, (for luck I presume), with Pink, White and Green[30] ribbon around its neck was given to the Colour Party, and I have kept the same.

We have been on the plains now for a fortnight and have had rain practically all the time. We have very good tent accommodation and have recently put wooden floors in the tents. There are some half dozen battalions encamped in this immediate vicinity, comprising about 10,000 men in all. Each crowd is to itself, the farthest apart being about three miles.

THURSDAY – NOVR. 5TH/14
Today I was placed in charge of the Transport (to go to Hdqtrs [Headquarters] with 3 or 4 men to get Rations for our Reg, for tomorrow). We got our transport from another Regt. & Ern[est] Gear [NI] was the driver. Am on this job for two or three days. Got back about 3 p.m. and was then free for the day. Worked at my old E.Coy pay book.

FRIDAY – NOVR. 6/14
On Transport work again today. Got back to Camp at 2 p.m. After dinner had to walk to Hdqtrs (4 mls) with Indent for Rations for Sunday, so as to draw Rations for Sunday as well as Sat. so as not to have to do so Sunday. Met a Witless Bay man today in a Mtl. [Montreal]Regt.

SATURDAY – NOVR. 7TH/14
Fine day today for the first time since being in Camp. Transport work again today. Finished the 2 trips at 4.30 p.m. Met one of the Jensens [NI] of Hr. Breton today. He was with the 5th Highlders of Mtl.[31]

SUNDAY – NOVR. 8TH/14
Went to Hdqtrs for veg[etables]. returned at 10 o'c. At 4 p.m. went with Capt. March[32] & Lt. Butler[33] to Lavington, had tea & went to Congl [Congregational]. Ch[aplain] Capt. Pringle [NI] – Chap. Can. Forces preached. Retd. to Camp at 9.30 p.m.

MONDAY – NOVR. 9/14
Went on transport & returned at 1.30 p.m. Nothing particular today.

TUESDAY NOVR. 10/14
Transport again today 5 p.m. – two loads.

WEDNESDAY – NOVR. 11/14
Did not go on transport today as I was needed in Camp – Bat.
Orderly Sergt. for the day.

THURSDAY – NOVR. 12/14
Sergt. of Guard. Nothing particular today.

FRIDAY NOVR. 13/14
On account of Reg. Sergt. Major [Regimental Sergeant Major]being
given leave – our Coy. Sergt. Major acted as R.S.M. and I was acting
as Coy. S.M. [Sergeant Major].
   At 8 o'c Bat. fell in for a test night attack. We went about 5 miles
and "took" or "captured" an old farm house. Returned to
Cp.[Camp] at 9.30.

SATURDAY – NOVR.14/14
Had special parade at 10 a.m. & marched to Black Heath for inspec-
tion by [Maj-]Genl. Alderson.[34] Reached back to Camp at 1.30 p.m.
Half holiday but spent day paying men of old E. Coy. Today acted as
Coy. Quartermaster Sergt.

SUNDAY – NOVR. 15/14
Held service for different denominations; those out of doors had to
be cut short as it commenced to rain heavily.
   Today owing to our acting C. Sgt having hurt his knee yesterday
playing football, I acted as C.S.M. as well as C.Q.M.S. [Company
Quartermaster Sergeant] was therefore busily engaged till 11 o'c.

MONDAY – NOVR. 16/14
Was very busy all day being C.S.M. & C.Q.M.S. We went on route
march this afternoon covering some 7 or 8 miles.

TUESDAY – NOVR. 17TH/14
Still C.S.M. & C.Q.M.S. and of course busy.

WEDNESDAY – NOVR. 18TH/14
Got uniform at last, this morning, and whilst walking across to tent
with it was asked to get 20 men to go to Lord Roberts[35] funeral. Left
Camp at 11 a.m. with Lt. Tait[36] in charge. Took train from Amesbury at 2 o'c & reached London at 6.15. Marched to Wellington Barracks for the night. Had a couple of hours leave so went to a restau-
rant for a good feed then had a stroll about.

THURSDAY – NOVR. 19TH/14
Left Barracks at 9 o'clock & marched to Charing Cross where we waited for some half hour or more, then moved on in the procession & eventually reached St. Paul's Cathedral where we halted for a short while, then went on and marched through various streets for an hour or more & then back to Barracks at 1 p.m. 2 p.m. left for Waterloo station & left for "home" by train at 2.45.

Left Amesbury about 6.30 by transport (motor) for Pond Farm.[37] When about 6 or 7 miles from destination motor gave out & could carry only a dozen, so a few of us had to walk. I took charge of the "walkers" (7 or 8) and we reached home about 9.30 p.m. – "All mud."

FRIDAY – NOVR. 20TH/14
About 10 o'c this a.m. Capt. Alexander[38] told me I had to go with him and 100 men to prepare for our Regiment moving to Bustard Camp.[39] We left at 3 p.m. & arrived at about 5 p.m. – Then had a very busy time until "lights out."

SATURDAY – NOVR. 20TH/14
Had very busy morning. Regiment arrived shortly after dinner time. All were kept busy during afternoon & evening.

SUNDAY – NOVR. 22ND/14
Bat. Ord. Sergt. today and was busy all day getting men to put up tents, marquees, and doing other work.

MONDAY – NOVR. 23RD/14
Nothing particular today.

TUESDAY – NOVR. 24TH/14
This evening after Tea we spent an hour and a half at "night work."

WEDNESDAY – NOVR. 25TH/14
Nothing beyond regular routine work today.

THURSDAY – NOVR. 26TH/14
Acted as Provost Sergt. today.

EXTRACT FROM LETTER DATED 26 NOVEMBER 1914.
The "Web Equipment," which we carry, is a multitudinous arrangement of straps and bags combining the following:– Belt, Bayonet Scabbard Carrier, right and left hand Cartridge Carriers, (capable of

carrying one hundred and fifty rounds), Haversack, Trenching Tool Carrier, Trenching Tool handle Carrier, Waterbottle, and Pack.

Last Friday I had to go with Capt. Alexander, one hundred men and another officer to Bustard to make arrangements for our Regiment to move there the following day. Needless to say, we had a very busy time, and I had to run nearly everything. The worst part of it was attending to supplying Rations. We really had enough for two meals only and I had to work it to do for three meals; the men were worrying the life out of me for food.

That night at Bustard was the coldest night we have experienced. The tent was all frost on the inside and our breath was frozen on our blankets, and I can tell you, it was pretty hard work getting out at six o'clock in the morning.

FRIDAY – NOVR. 27TH/14
We were Duty Battalion today. I was in charge of 3 Sgts & 2 police from our Reg. and some dozen Police of other Regiments on Canteen duty.

SATURDAY – NOVR. 28TH/14
Heard today that Ed. Murray [NI] was up to Camp yesterday but I did not see him owing to being away on duty.

Today Uncle William[40] came to Camp. He brot a dozen cakes – 2 for me. He went back to Salisbury for the night.

SUNDAY – NOVR. 29TH/14
Uncle Will came back just after we had returned from Church in the Y.M.C.A. tent. He brot a pound box of choc. for me and stayed until middle of afternoon. Having leave I then walked to W. Levington & returned at midnight.

MONDAY – NOVR. 30TH/14
Nothing unusual today.

TUESDAY – DEC. 1ST/14
Very little done today owing to constant rain.

WEDNESDAY – DEC. 2ND/14
Went on a seven or eight mile route march in the afternoon.

EXTRACTS FROM LETTER DATED 2 DECEMBER 1914
When in St. John's, we would often complain of the mud, but after the mud we have experienced since our arrival on Salisbury Plains, I

do not think that any of us will complain or even think unkindly of any mud we may encounter in our home city. We have not only seen more mud than we have ever seen before, but we have walked in it nearly every hour of the day. Many of us, particularly our transport men, have been compelled to walk in mud that was over the tops of our boots, which has given us lots of work in trying to clean our boots and puttees again, and I have been in such mud as this, time and time again.

By the way, there is a farm about half a mile from our present camp bearing the name "Newfoundland Farm," and it is designated as such on the Salisbury Plains' Maps. How it came to bear this name, I have not been able to find out, but it is certainly a co-incidence that we, as Newfoundlanders, should be located so near to it.

All around our present camping ground are fully three thousand tents, whilst as many more are visible within a mile or two of us. When one realises that these tents are the present habitations of thousands and thousands of men, all busily and earnestly under going strenuous training with the object of fighting for the honour, and may be, the preservation of our great nation, one then realises also what a very serious undertaking is being carried on.

The system of training consists of Physical Exercises, Marching, Rifle Exercises, Skirmishing, Route Marches &c.

We are all very particular here that we not be classed as Canadians, for apart from the fact that we are much prouder of our distinction as Newfoundlanders, the Canadians, generally, have been getting a bad name for themselves wherever they have gone, and this applies particularly to London. This is very unfortunate for them, for when their first Draft came across the water, the English people could not do enough for them and London was strongly agitating for them to be brought there to be seen. On the other hand, I am pleased to state that the Newfoundlanders have been given a very good name wherever they have gone, and then our general excellence has been of a very high standard, certainly much higher than that of the Canadians. We have received the highest praise from several Colonels and Generals, who have said that we were one of the finest bodies of troops on the Plains. Our greatest regret is that we are only half a battalion, and the Headquarters Staff desired to attach us to a certain half battalion of Canadians, now encamped alongside of us, but I am pleased to say our officers are very strongly opposing any such intention. It is rumoured that we may be attached to some English Regiment, that is, if we do not soon hear that Newfoundland is prepared to send over a second contingent of five hundred and then give us our own full battalion.

For the majority here, it is not all hard work, for there is a "Movie" here which is well patronised every night. Then there is a Y.M.C.A. Tent which also has its capacity tested to its utmost every night. About one-third of this tent, which is about two hundred and fifty feet by fifty feet, is occupied by tables which are crowded every night with men writing letters. The remainder of this tent is used as a concert hall and there is a piano. An impromptu concert is held here every night where the men have the pleasure of listening to songs, recitations &c., and even in this diversion from the daily routine of a training camp, our boys are always well to the front and contribute their share to the programme.

A few days or rather nights ago, a very amusing incident occurred. Sentries are posted on specified beats and after "lights out" (9.45 p.m.). They are supposed to "Halt" everyone whom they see, and ascertain if they are friends or otherwise. One of our Sentries was doing duty on his allotted beat, when he thought he saw two men moving about a short distance off. As the sentry was pacing about, he shouted to these objects to "halt," but they apparently took no notice; he therefore shouted again with a like result; after shouting a third time, and the figures still appearing to move, he called for the whole guard to be brought out which consisted of about fifteen or twenty men, who turned out with rifles and fixed bayonets and upon investigation the two figures turned out to be two goal posts. The seeming movements of the two figures, or posts, were caused by the movements of the sentry himself. Needless to say, the guard men were disgusted and immediately christened the sentry "Goal Post" which name still sticks to him.

THURSDAY – DEC.3RD/14

Inspection by Genl. Alderson in the morning when he told us that the War Office had just informed him that we're shortly to be sent to Barracks at Fort George – (near Inverness Scot.) & that our battalion was going to be increased to a full Battalion of Newfoundlanders.

After dinner we marched to another place two miles away to be inspected by Lord Brass[e]y.[41] He afterwards made us a short speech.. We returned to Camp about 2.30 p.m. and were freed for the day.

EXTRACTS FROM LETTER DATED 3 DECEMBER 1914.

To-day has been a Red Letter Day for our Contingent. First of all, it is exactly two months today since we left Pleasantville, and when thinking over the events of that period, I must confess that it seems considerably more than two months since we camped on the grounds at Pleasantville, previous to our march through the City to the "*Florizel.*"

This morning our first Parade was an Inspection Parade, and we 'fell in' in full marching order, with full equipment. We were inspected by General Alderson, who is in charge of the whole of the Canadian Forces on the Plains. Before inspecting us he had us drawn up in close order and told us that he had just received word from the War Office that it had been decided to have us sent to – and at this point many of us saw pictures of the front, thinking his next words were going to be "The Front," but what he did say was "Barracks." However, this was very pleasant news for us, for the majority of troops have to move into huts or shacks, and it is considered a "clinch"[42] and a privilege to be put in barracks, for a Regiment must have a very good record indeed to get there. This decision on the part of the authorities, certainly speaks well for the behaviour of our Regiment and the high standard it is fast attaining. The General went on to say that he had not been informed as to just when we were going to be moved, but that we were going to Fort George. Now there are at least four Fort Georges, – one near Portsmouth, one at Inverness, Scotland, one at Seaforth near Glasgow, and one in Ireland. No one knows just to which we are going, some think to Portsmouth, but there is a strong rumour in favour of Inverness. The General also said that he was glad to hear that it was the intention to convert us into the strength of a full Battalion of Newfoundlanders, as this would permit us to preserve our identity; and that if he were a Newfoundlander, he would certainly wish for the same. He said that we were to continue our training at Fort George under more favourable conditions until our second Contingent joined us.

We were then opened out for inspection, and after inspecting us, General Alderson said that he was very pleased with our general smart appearance. We then performed some skirmishing work for his observation, and perhaps criticism. After that we returned to camp for Dinner, and later "fell in" for a second inspection. This time we had to march a couple of miles and were inspected by Lord Brass[e]y. It will be remembered that Lord Brass[e]y paid an official visit to Newfoundland a couple of years ago. After inspecting us, Lord Brass[e]y spoke to us, saying that he was asked by our Government if he would be a member of the Newfoundland War Contingent Association and assist in the reception and welcoming of the Newfoundlanders to their Mother Country. England, and the English people, he stated, were pleased and proud of us, and it afforded him great pleasure in welcoming us to these shores. He said he knew we would do credit to ourselves and our country, and that we would eventually go where honour and glory awaits us.

Continuing, he said that it was intended to give us a reception of some sort or other, but he could do nothing in this matter until the General had been conferred with, to see in what manner this could be carried out. There had, as yet, been no meeting of the Association, and the time for their usefulness had not yet come. In concluding, he said that it was his intention to write our Government, informing it, that he had been down to see us, (as this was about all he could present), and that he would express to them, the pleasure he had had in doing so, and how pleased he was at our appearance. We then returned to Camp, and were afterwards dismissed for the afternoon.

FRIDAY – DEC. 4TH/14
After this morning's parade I secured my pass but with great difficulty. Left Camp in rainstorm by motor car with L. Bain[e]⁴³ for Devizes at 12.30 p.m. arrived there at 2 p.m. and half hour later took train for London, arriving there at 6.30 p.m. Finding that a train left for Gravesend at 7 p.m., I telegraphed Aunt Nellie I was coming on that train and then joined that train. Arrived at Gravesend about 8 p.m. & Aunt N. had just reached the station to meet me. Just after reaching home Mr. L.[Baine] came home. Both greatly surprised. Had good bath tonight and then a good sleep.

SATURDAY – DEC. 5TH/14
After breakfast the 3 of us went for a couple of hours walk & then Aunt N. & I did a little shopping. After dinner Aunt & I went to a "Movie" & did a little more shopping. We stayed in after tea chatting. I left by train for London at 10.30 p.m. Aunt coming to station with me, Uncle having had to go off to a steamer. Reached London 11.30 p.m. & went by tube to St. Pancras intending to take train 4.50 a.m. When I got to station found no train leaving at 4.50 owing to being Sunday. There was one leaving at 12.05 and it was just then 12 o'c – so I booked to Greetland (near Halifax) [in Yorkshire] and caught train.

SUNDAY – DEC. 6TH/14
Left London 12.05 a.m. & arrived Normanton 5.20. Stayed in waiting room. Left for Gld. [Greetland] at 9.20 & arrived Gld. 10.20. Met by Uncle [Frank Bottomley] & walked home (4 hrs). Spent afternoon & evening sitting around chatting & with the gramophone. Thought Daisy & Billie were nice girls, particularly B. – Liked Uncle & Aunt [Laura Bottomley]⁴⁴ very much indeed.
    [Note: No entries as Steele is on furlough]

SATURDAY – DEC. 12TH/14
Warrant came by postman at 9 a.m. so immediately prepared for departure.

[Note: The following appears only in the original, not in the type-script copy]
At 10.15 I said goodbye to Aunt & Daisy & left Bowers Hall with Billie to go to Gld. for my other suitcase. We then left by train to Halifax & went to Uncle's warehouse. He then came across to the station and arranged about my ticket. Arriving back to warehouse I said goodbye to him and old Mr. B[ottomley]. Billie & I then did a little shopping & then went to the Imperial Café and had lunch.

Left for North at 1.50 , Billie seeing me off at station. Arrived at Keighley 2.40 & changed leaving at 3.28. Reached Hellifield 4 o'c & changed leaving at 4.30. Arrived Edinburgh 8.50 & unable to leave until 7.35 a.m. Sunday. Took a room at the Cockburn Hotel and then went for a walk. Weather raw & cold & rain falling. Too wet to be enjoyable. (Wrote P.C.s [postcards] to Home, to Aunt Laura & to A.B. Young[45] – Kirkcaldy [in Fife, Scotland].

Had snow as soon as we entered Scotland.

SUNDAY – DEC. 13TH/14
Got up at 6.45 a.m. Was on board train at 7.30 but did not leave until 8.30 owing to London train being late caused by engine of another train breaking down.

Arrived at Perth at 9.45 a.m. changed trains and & left again at 10 o'c. Arrived at Gollanfield junct. at 2.15 & was given a motor ride by an officer (Capt.) to Fort George arriving 3 p.m.

[Note: The typescript contains the following additional detail, pre-sumably taken from a letter.]
Fort George is an old Fort which was built during the reign of George II, in the year 1757. It is said that the architect, after seeing the Fort when finished, committed suicide. The Fort is not now forti-fied, owing, I presume, to the fact that Cromarty, which is only about ten miles away on the opposite side of Moray Firth, is one of the East Coast Naval Bases, and is therefore strongly fortified. There is a large number of all kinds of warships there, and the sky and surrounding sea and land are brilliantly lit up each night with search-lights, watching for the approach of any German submarines, ships, or zeppelins.

CHAPTER TWO

Scotland
(13 December 1914 – 2 August 1915)

FORT GEORGE
(13 DECEMBER 1914–18 FEBRUARY 1915)[1]

The regiment was in Scotland for nearly eight months, its stay there broken up by frequent moves that seemed to hold out little promise of immediate action. Nevertheless the time spent in Scotland was, for the most part, pleasant, if at times monotonous.

The first location was at Fort George in northern Scotland. Fort George, built in 1748, three years after the failure of the Jacobite Rebellion, was reputed to be a fortress of considerable strength and ideally situated on high ground dominating the western approaches to the Mornay Firth. The Firth was the location of a seaplane base that provided aerial reconnaissance for nearby Cromarty, one of the east coast naval bases, and kept watch against German Zeppelins. One of the duties assigned to the Regiment was to provide the nightly guard for the seaplane base. Fort George had been the depot for the famous Seaforth Highlanders until they went to the front after which it became the training centre for large numbers of recruits for Kitchener's New Army. The Newfoundlanders were the first colonial troops to be trained at Fort George and because of this created considerable curiosity amongst the locals.[2]

The Newfoundlanders liked Fort George and found garrison life more agreeable than camp life on Salisbury Plains. The routine was not onerous and included drill, musketry, [rifle firing] and frequent route marches intended to strengthen bodies and harden up the feet. Weekend leaves were generally spent in Inverness, within easy reach of the fort. The countryside around Fort George reminded Steele of home, in particular the rugged coastline of Fortune Bay. In these more familiar surroundings the men of the regiment began to develop close bonds with one another. Steele certainly became more aware of the links between the families of men in the regiment and his own in distant Newfoundland.

Fort George provided Steele with a further occasion to prove his worth as a capable NCO. His appointment as regimental provost marshal, responsible for the preservation of order in and out of barracks, and his command of a small detachment indicated a high level of trust in his capabilities and good judgement. Steele welcomed the opportunity and hoped it might open up the possibility of a recommendation for a commission, something he clearly desired.

By February 1915 Steele's initial enthusiasm for Fort George and garrison duty had worn off. Christmas parcels from home and weekend excursions brought some brief diversion but did little to alleviate the increasing boredom. His letters, the only record we have of his time in Fort George, reflect a growing tedium and increasing concern that the regiment might spend the entire war in garrison duty. Until C Company arrived the regiment was still under strength and unlikely to go anywhere and Steele knew this. The front was still some distant place – alluring but beyond comprehension. Steele had a brief encounter with a wounded sergeant from the First Scots Guards whose exploits and DCM served only to whet his appetite for adventure. Finally, news of the departure of the second contingent from St John's raised spirits and held out the promise of a move, as it turned out, not to the front but to historic Edinburgh.

EXTRACT FROM LETTER DATED 25 DECEMBER 1914

I have decided to write once every week, at least, and if I have not the time to write a letter when the time to write comes, I shall send a Post Card, – I shall write every Sunday.

I enjoyed my holidays immensely and shall send full particulars tomorrow or the next day. I reached Barracks on Dec. 13th, and the next day was appointed Provost Sergeant. Another chap had held this position but he was not satisfactory, and I was recommended for it by Capt. Alexander and the Garrison Sergeant Major. Colonel [R. de H]

Burton[3] sent for me on Monday, and said he was thinking of making
a change in the Provost Sergeant and asked if I would like to take it. I
then asked if it were permanent or temporary, for if permanent I
would not like to take it. He looked at me once or twice, then at his
desk and then said, "Well we'll make it temporary." So I took it. It is
not a promotion but simply an appointment, and is usually held by a
Sergeant. The Provost Sergeant is responsible for the preservance of
order, in and out of Barracks, and has, as assistants, a number of men
acting as policemen really, (I have four); the Provost is really a
Sergeant of Police. My men are Frank Knight,[4] (son of Herbert Knight
Esq.), Vail,[5] (who used to be a policeman in St. John's), Mifflin[6] (from
Bonavista and a relative of Heber Mifflin [NI]), and LeGrow,[7] (a relative of LeGrows of Broad Cove, Bay de Verde). The other Provost and
his men were under the control of our Garrison Sergeant Major, (G.
Paver[8]), but when I was appointed, the Colonel said that everything
was to be in my charge and under my control.

A few days ago, I went to Nairn by train, (about twelve miles by
rail), to locate and bring in one of our Sergeants. Military Regulations require that a Sergeant must be apprehended by some one
above him in Rank, even a Sergeant cannot be sent after him. All the
Sergeants laughed at the idea of my going after him and freely said
that they were positive I would not be able to get him back. Nevertheless, to make a long story short, I brought him back (myself), and
what was more brought back two other men besides. Capt. Alexander told me I was lucky to get this position for it would give me an
opportunity to show myself up before the Colonel for a Commission.
I would not be bothered with Commission in an outside Regiment,
for I wish to remain with our own crowd.

EXTRACT FROM A LETTER DATED 1 JANUARY 1915
I have succeeded in getting "New Year Holidays." I did not expect
any more than three, or at the most, four days, but when asked how
many I wanted, I said, "As long as you can let me have." So they
said, or rather asked, "How would six or seven days do?" Of course,
I said that that would be all right. I left Fort George Barracks at
10.45 a.m. Thursday, and arrived in Halifax 7.15 this morning, after
a very weary and tiresome journey, during which time I had no sleep
whatever, having had to keep changing trains. When coming down
through Scotland, as we drew near the Forth Bridge, an Inspector
came along and had all our luggage taken from us and put into a special van, in case, I presume, anyone had anything to damage the
bridge or Navy boats in its vicinity. Then, of course, blinds have to
be kept drawn at all times.

We all like the Fort life very well indeed. The country round about is very like Newfoundland, and sometimes I can fancy myself in parts of Fortune Bay. I shall send you a few Post Cards of the fort as soon as I can get then – they are getting a stock in shortly.

We are divided into rooms of five men each, in the main buildings of the Fort, and in the Casements of about a dozen men each. The Casements are long passages built into the walls of the forts and are considered bomb-proof; there are several feet of earth on top of them. I am in one of the 'five men' rooms. There is only one in my room whom you would know, that is Frank Knight.

We do not have to get up so early in the mornings now. Reveille does not sound until 7 o'clock. Breakfast is at 8 o'clock; first Parade is from 9.15 to 12 noon, then Dinner; Parade again from 2 o'clock to 4 o'clock, Tea 5 p.m., First Post 9.30 p.m., Last Post 10 p. m., Lights Out 10.15 p.m. First Post is sounded on the bugle for lads to get to their own quarters where the Roll is to be called. Last Post is sounded to warn them that "Lights out" will be sounded fifteen minutes later. The Parades vary, sometimes being for Drill, sometimes for Musketry Exercises and sometimes for a route march. As a matter of fact we are all having as good a time as could be expected under the circumstances. We can get baths wherever we like, but it is cold water of course. I do not know whether it will be possible to get warm water later on, but it is very hard to clean oneself with cold water.

Alongside the Fort is an Air-ship shed and in it are four Air-ships. I have seen them go up and come down several times. I should have said they were Hydroplanes for they are both Air and Water ships.

EXTRACT FROM LETTER DATED 12 JANUARY 1915.
This is Tuesday, and I should, according to my promised intention, have written on Sunday, but I got back from my holidays only last night.

At first, let me say that on my return I got my Christmas Box from Home (for which I thank you very much indeed), containing all the nice things, luxuries and useful articles. I shall certainly enjoy them, for the Cakes &c. will come in very nicely for suppers. Am sorry to say the poor bottle of molasses got broken during the journey and the "rare treat" as mother marked on it, found its way even to the bottom of the tin can containing the Partridge, and Mince Pies, but you had fortunately wrapped each Pie and Cake up individually; everything was O.K. Of course, after such a long journey, and probably lying here for a week before I got it, owing to my being away on leave, the poor partridge got a little mouldy, but I am going to try it for supper to-night.

This week is half gone, so I may not be writing the coming Sunday, for I have quite a little correspondence to attend to in connection with Xmas presents.

EXTRACT OF LETTER DATED 31 JANUARY 1915.
We are all anxiously waiting to hear of the departure of the Second Contingent. I, for one, do hope they will be enabled to despatch 500 men, so as to make us a full battalion.

A few evenings ago matters were quite interesting here, for on Wednesday evening a German Zeppelin was seen eight miles off Cromarty, the great Naval Base. At 9 p.m. all lights in the Fort were ordered to be put out; all guards were doubled and other precautions for an attack were taken. The doors of the Aeroplane sheds, about which I mentioned before, were opened and the planes were taken half way out and arrangements made for instant ascent. The reason the Germans are now after this district is because Cromarty is a great Naval Base and there is always a large number of warships there; and then there is the Fort here with over fifteen hundred men, and of greater importance, the Aeroplanes and sheds. Nothing has since occurred, but we are all on the alert as it is said that the Germans are going to keep trying until they get here. It is fine at nights to see the Search Lights from the Warships on guard at Cromarty, about five or six of them, searching the surrounding country, the sea, and sky, looking for suspicious movements. It will certainly be interesting for all concerned if they do come, as we are all anxious to get into action

EXTRACT FROM LETTER DATED 9 FEBRUARY 1915.
I must commence this week's letter with an apology, for I did not write on Sunday, this being, as you will see from above date, Tuesday.

For the past fortnight 'B' Company, to which I am attached, has been having its turn at Musketry Practices, and all belonging to 'B' Company had to be on parade; those doing any special duties, such as Pioneers, Cooks, Guards, &c. were relieved of their duties by 'A' Company. I was relieved of the 'Provostship' so as to be able to take part in our Musketry, and we finished same on Saturday, so we of 'B' Company have again taken up our Duties.

A lot of our boys now get weekend leave, (from Saturday afternoon to Sunday night or Monday morn ing), so I got permission to go to Inverness for Sunday. I went Saturday evening just after tea and came back Sunday night. It is quite a change from the monotony in the Fort. – Hence the reason I did not write on Sunday; then Monday (yesterday) I did not get time.

I went to church both morning and night on Sunday. In the morning we (Sergeant Malcolm Godden,[9] son of Mr. Godden at Dicks and Coy., was with me) went to the United Free Church, and, as we were a little early, stayed in the vestibule for a few minutes talking to one of the ushers, and just as we had decided to go in and sit down, he said that the service was to be conducted in Gaelic that morning, so we were not long in "beating it" elsewhere, and went to another Presbyterian Church. Malcolm Godden is now Orderly Room Sergeant and holds the position that Walter Rendell[10] wanted me to take when at Pleasantville.

Inverness is not a bad little place, being about two-thirds the size of St. John's, and contains very pretty scenery on the outskirts, and reminds one very forcibly of Bowring Park.

It is reported that the Second Contingent is now on the way over, having left St. John's by the *Mongolian*[11] on February 6th.

EXTRACT OF LETTER DATED 15 FEBRUARY 1915.

Well, I believe, I am writing you for the last time from Fort George, for on Saturday, the 13th inst., orders came through to the Regiment, that we are to go to Edinburgh. At present it is understood that we are going on Thursday, the 18th inst., and that we are going to be quartered in the castle. The Second Contingent is going there direct from Glasgow and are to be there to-day, Monday. Our Advance party went out yesterday morning. Some reports say that we shall be going to the front in a fortnight's time, but I am afraid we shall not be going for some time yet, though I do hope we shall soon have that pleasure, for we came over for that purpose, not to spend all our time in Barracks &c., and would be almost ashamed to go home again if we do not get to the front first.

I spent yesterday afternoon and evening at Mr. and Mrs. D. Petries', Inverness, (having received an invitation a few weeks ago per Capt. March), Mrs. Sclater and Mrs. Miller having given her my name. Mrs. Petries said that Mrs. Slater and Mrs. Miller were very disappointed that they did not see me when they were visiting the Fort, – I was away somewhere then. Just after we had finished tea, Capt. Rendell and Lieut. Butler came in. They stayed until shortly after six o'clock. We spent the afternoon chatting and with music. They are quite a musical family. Mr. Petries sings, plays the Cello and the piano; Mrs. Petries sings; one of their daughters sings, plays the piano and the violin; another sings and plays the piano; the third, unfortunately, is devoid of musical taste. Mr. Petries and one of his daughters are in the choir, singing Tenor and Alto respectively, so asked me to come in the choir with them, and as the Basses were their weakest part, I had my own way pretty well. We sang "Lord for Thy Tender Mercies Sake," and Jackson's "Te Deum" both of which I knew well, and, of course, the hymns were no trouble. When we returned from Church we had supper and then family Prayers were held, after which we had quite an enjoyable time with anthems and quartettes [sic], the latter including "The Heavens are telling," "O taste and see," "The Radiant Morn," "Abide with me," and "Sun of my Soul." We also had my favourite hymn "Dear Lord and Father of Mankind." The Petries are Presbyterians, though Mrs. Petrie was originally a Congregationalist.

A Sergeant Shields of the First Scots Guards was also visiting them in the afternoon. He has only recently returned from the front, having come home wounded. He was standing in his trench with apparently no live Germans anywhere near, but a few were lying about six yards in front of the trench, all of whom they thought were dead. However, one of them was only wounded and got his rifle into position and fired at Shields. The bullet went into his face just under his left eye and passed through behind his nose, and came out on the right side midway between his cheek bone and right ear. He did not know it had happened until blood came down his nose. He said that after it happened he fired about one hundred and fifty more rounds.

He has won the Distinguished Conduct Medal. He crawled from his own trench on his stomach to behind the German trenches, a half mile away, to get some information. He found out how many men were there and other valuable information, and was starting back when he was challenged by a German Sentry, whom he immediately shot down and then took to his heels for all he was worth. A fusilade of shots came after him, but none touched him and he got back safely. Another time he found that some of the men in his trench were being picked off by snipers on the right flank, so he made a detour and got behind them and shot the lot, fifteen in all. When he returned he told his officer, who said it seemed a very tall story to him and which he could not believe until he had seen it. Shields told him to come over and see for himself. They counted fourteen dead, but could not find the fifteenth.

EDINBURGH (19 FEBRUARY – 11 MAY 1915)

The regimental historian recorded that the Newfoundlanders received a tumultuous welcome in Edinburgh, the first "colonials" to be stationed in the ancient capital.[12] Their arrival was preceded by the much-awaited second contingent, strangely attired in their Newfoundland-made uniforms. The resulting excitement of meeting old friends and catching up on news from home brightened up Steele's letters and gave him new material to write about. So, too, did the delights of life in a large centre, notably the theatre and the sights of the historic city which he shared on occasion with his brother Jim. A steady stream of visiting Newfoundlanders, some known to his family, allowed him to write about something familiar.

Steele continued his duties as Provost Sergeant, in which he enjoyed considerable success. Evidently he still had his eye on a commission. He was hardly able to contain his disappointment when fellow sergeant Jack Fox was singled out for commissioning because of his family's political connections. When his own turn came a few weeks later he clearly felt suitably rewarded and visibly pleased by all the accompanying attention.

It was while at Edinburgh that Steele received the sad news of the death of his younger brother Herbert, whose promising career as a Baltimore dentist had been cut short by tuberculosis. The diary is surprisingly silent on Steele's thoughts about Herbert's death and he mentions only that he dropped out of the regimental cross country race when he received the news.

By mid April rumours abounded of an impending move, possibly to Egypt, but the eastern Mediterranean was still months away. Anxious as ever for some action, the Newfoundlanders had to make do with public approbation for their turnout in a monster march in April and with an official visit by Sir Edgar Bowring, the colony's high commissioner in London.

EXTRACT OF LETTER DATED 21 FEBRUARY 1915

We are now quartered in the Castle, having arrived here the night before last, – Friday – at 7 p.m. I expect you have received my card sent from Fort George just as we were leaving. The Second Contingent was already here when we arrived, having arrived on Tuesday evening, three days previously. I have not yet had time to get around to see things properly but have come across quite a few of those whom I know.

The Second Contingent has not yet got their Regimental caps and are still wearing those woollen caps, but their uniforms are very good for St. John's make. One thing there is sure, they, of the Second Contingent, have certainly got the soft end of the plank. They came across in a comfortable steamer and have come straight to Edinburgh Castle, not having to undergo any of the really severe hardships we underwent on Salisbury Plains for seven weeks. Everyone agrees that we should have been kept at Fort George and that the Second Contingent should have been sent up there to get through their training. Facilities, as far as we can see at present, were far better for training purposes, and those of the Second Contingent thought they were going there until they disembarked in Edinburgh.

I have often heard Scotch people speak of the beauty of their Highland Scenery and always thought it the result of prejudice, but in future I shall always agree that it is about as fine as anyone would wish to see, for our journey from Fort George on Friday was about the finest sight-seeing trip I have yet seen. There was practically no snow within a few miles of either side of the railroad. The ground was fairly level and in its natural colour of various shades. At the back of this were high mountains of innumerable shapes all covered with snow. It was a gloriously fine day, and the sun was shining in all his glory, showing the snow-clad mountains up splendidly. The scenes were of the greatest variety and constantly changing. Of course, portions of scenery in the Canadian Rocky's surpasses that seen on the Highland Railway, but it is much more general throughout the trip. I only wish you could have been here to see it, for I am sure you would have enjoyed it. I know I was quite enraptured.

EXTRACT FROM LETTER DATED 27 FEBRUARY 1915.
This afternoon I am playing in a Rugby Match against a college team here. There is a notice posted up saying "The Colonel has kindly consented that the following will represent the Regiment in Rugby match against —." There are at least four officers playing, probably five, namely, Dr. Wakefield and Lieutenants Tait, Raley,[13] Alderdyce[14] and perhaps Wighton.[15] This will be my first match. A couple of matches were played at Inverness when we were at the Fort.

EXTRACT FROM LETTER DATED 28 FEBRUARY 1915 (CONT.).
I did not play in the Rugby match yesterday, for just when ready to leave the Barracks, the Colonel sent for me and the Sergeant Major (Regimental) regarding certain Barrack work, and when I got to the University Grounds, the first half was over, so it was not worth while for me to begin then. Our side played two men short, (thirteen instead of fifteen) as another chap did not turn up, and was consequently badly beaten.

On Friday night, I went to the theatre with Frank Knight, for Gladys Cooper[16] and Seymour Hicks[17] were giving a sketch, being a week's engagement. Then last night we went to another place to hear Martin Harvey,[18] one of the "Stars" of the day, – he is shortly to be knighted they say. Gladys Cooper and Seymour Hicks were quite good but not extra, but Martin Harvey was really wonderful.

Everything was so real and the stage settings were perfect. We were entranced almost, the whole time. Geo. Langmead[19] sat two seats in front of us.

To-day Charlie Marshall [NI] was here with Nat Snow [NI] and I was speaking to them. John Jackson was also here with Miss May Furlong [NI], but I did not see them.

Dr. [Lamont] Patterson[20][Paterson] and Capt. Montgomery[21] [Montgomerie], also George Kearney and Donald Nicholson [NI], were also here, all of whom I have been speaking to.

Next Tuesday five of our officers, namely, Capt. O'Brien,[22] Lieutenants Wighton, Ayre,[23] Alderdyce and Nunns,[24] are returning to Newfoundland. The first three are going to assist in bringing over the 3rd Contingent, and the other two are remaining in St. John's to assist in the training of the reserves. W. Grant,[25] whose father died yesterday or to-day, is going back also; he goes as Capt. O'Brien's orderly and returns again with the 3rd Contingent.

I have promised to spend the weekend of March 27th with Mr. & Mrs. Young of Kirkcaldy.

Have seen a little of the sights of interest in Edinburgh, but hope soon to be able to do so more extensively. I want to visit Holyrood Palace, The National Gallery, The Museum, The Zoo &c.

We have been having lovely weather ever since our arrival here, having had a fall of snow, of about half an inch, a day.

I am still Provost, being now Provost of Edinburgh Castle instead of Fort George.

EXTRACT FROM LETTER DATED 15 MARCH 1915.
One day last week, about 9 o'clock in the evening, I had to go to Leith, – a suburban district, – with two of our Regimental Police, for the purpose of bringing in a Deserter from the 4th Battalion Highland Light Infantry. He had been absent ever since December and his Battalion (now in Bristol I think) had heard that he was at a certain address and so wired the Castle to endeavour to apprehend him. So I was sent by the Colonel and the Adjutant, and was told that it was quite a great thing for us to be asked to do this and that it was quite a responsible undertaking for me. The Adjutant told me to report to him as soon as I got back, whether before or after midnight, and if I had the man or not. However, I got back with the man just before midnight, having found him in bed; he was about 46 years of age.

Several Newfoundlanders have been here lately, Hugh Anderson [NI], A.K. Lumsden [NI], J. R. Stick,[26] Miss Stick [NI], Miss Seymour [NI], and Ken Blair [NI].

EXTRACT FROM LETTER DATED 6 APRIL 1915

We have just been informed through our Regimental Daily Orders, that the appointment to a commission of Sergeant J.E.J. Fox[27] has been approved by the Military Secretary. No one begrudges him his Commission because he has stuck to his work well and is a good man; also on account of his connection with our Premier, Sir. E. P. Morris.[28]

EXTRACT FROM LETTER DATED 11 APRIL 1915

A short time ago there was a Harrier's Cross Country run of seven miles, just outside Edinburgh, and our Regiment was asked to send in a team of twelve men. The first six men of each team to finish were to count for points. The race was to take place on March 20th, and we were asked about March 1st. I was asked if I would undertake to get a team together and train them, so I agreed. In two or three days I had a team, varying from twenty to twenty-five men, practicing daily. I had a regular training schedule drawn up, and made great progress and had some very good runners. Then I got news of Bert's [his brother] death on March 16th, four days prior to the race, so dropped out. I would have done very well in that race for I used to win all the practices. There were about two hundred and fifty in the race, and out of about twenty teams, I believe our team got sixth place. Our first man to finish was Stan Greene [Green][29] who has won several six mile races in St. John's, and has done well in the ten and fifteen mile races. He got tenth place.

The Military at present quartered in and around Edinburgh, took part this afternoon in a monster March Out, the largest ever seen in the Capital. The number of men in the ranks exceeded 12,400, and the column extended three and a half miles. The muster ground was the meadows.

The order of March was as follows:

ORDER OF THE MARCH.

2/1st Lothians and Border Horse.

Composite Brigade, under Colonel Sir R. Cranston, K.C.V.O., C.B., V.D.

The Harry Lauder Band.

15th Royal Scots Depot.

16th Royal Scots.

NEWFOUNDLAND REGIMENT.

Hants Brigade Prov. Battalion.

1st Lowland (R.F.A.) Brigade.

4th London (Howitzer) Brigade, R.F.A.

Lothian Infantry Brigade, under Brigadier General H.F. Kays.

1|4th Battalion The Royal Scots.
2/5th Battalion The Royal Scots.
2/6th Battalion The Royal Scots.
1/7th Battalion The Royal Scots
2/9th Battalion The Royal Scots.
2/1st Lowland Mounted Brigade T. and S. Column, A.S.C.
3rd Lowland Field Ambulance, R.A.M.C.

All along the route crowds of spectators lined the way, and at certain points such as Tollcross, the West End, the Mound, the Post Office, Abbeymount, and the High Street, the congestion was pronounced. Everywhere the deepest interest was manifested in the spectacle, and perhaps the units most eagerly commented on were the NEWFOUNDLANDERS. (*Edinburgh Evening Times*).

With a view to translating the interest aroused into action, recruiting meetings were held immediately afterwards at Maule's Corner and Tollcross.

EXTRACT FROM LETTER DATED 19 APRIL 1915
I have been through Holyrood Palace, the Zoo, and the Camera Obscura Building or the Outlook Tower. They were very interesting indeed and also instructive. In the last named place there are many old sketches, engravings &c. of Edinburgh, and many other things of interest. In the Tower, by means of a Periscope, one can see the various portions of Edinburgh on a canvas table. By moving a handle one can see any part desired and can see the people walking about. It is certainly very interesting. Jim [his brother] has been through the Palace and the Zoo, but not the Outlook Tower. We have yet to visit the National Art Gallery and the Museum. On Sunday past we went in a Bus to Forth Bridge, for Jim wished to see it. It was of course nothing new to me, for I have been over it six times, each time in daylight. The run in the Bus took us forty minutes and the fare is 1/- per head each way. We saw some fairly good rural scenery on the way. The various approaches to the Bridge are guarded by Sentries, and no one is allowed to pass them in the direction of the Bridge without a pass, except soldiers in uniform.

There are some rumours of an early move for us, but we do not know where; some say to Egypt, but no one knows for sure when or where.

EXTRACT FROM LETTER DATED 23 APRIL 1915
Well! I have to-day started a new stage of my upward climb, having received a Commission in the First Newfoundland Regiment. The

Colonel welcomed me to the Mess at lunch time, for I happened to sit directly opposite him. As an officer, one has one annoyance and that is saluting, – I am sick of it already. I was vaccinated yesterday, four spots to make sure.

I feel pretty sure we are to move within a week but do not know where.

EXTRACT FROM LETTER DATED 3 MAY 1915

Yesterday, Sunday, Sir Edgar Bowring, Henry Bowring Esq.,[30] and his two daughters were with us for lunch, and Sir Edgar, by letter a day or two before, invited all the officers to Dinner at the Caledonian Hotel, for to-night, Monday. Unfortunately, my turn for Orderly Officer has come around to-day, so I am out of it. It is now 10 p.m. and am not yet finished my day's work, for about 11 or 11.30 p.m. I have to inspect the guard. The Colonel said that one officer must remain with me in the Barracks, so Lieutenant Rowsell said he would stay. But at the last moment to-night, about 7.30, word came that the other Company, (E) was coming and expected to arrive at 9.30 p.m., so the Adjutant (Walter Rendell), Quartermaster (Lieut. Summers[31]), Transport Officer (Lieut. Raley) and the Colonel had to remain from the Dinner. Word has since come that E. Company will not be here until midnight, or may not be here until sometime to-morrow, so we do not know what to do. If they come before 6 o'clock tomorrow morning, I shall have to be around, being Orderly Officer.

I have just returned again after having been away for two hours, for it is now twenty minutes to two. The Colonel sent for me, – for the Orderly Officer is everything – the all in all – in Barracks.

We now find that the Company will not be here until 3 o'clock, but we shall not go to sleep, for, as they say, we shall be too sleepy to know what we are doing when 3 o'clock comes, and besides I may be sent for at any minute.

Have just returned from the Station, the second crowd having arrived at 2.30 a.m. It is now quarter to four.

STOB'S CAMP: (11 MAY 1915 – 2 AUGUST 1915)

Stobs Camp, 90 kms South-East of Edinburgh, was a small Scottish outpost of 20,000 in an area noted for its tweeds. The closest community was Hawick, described by Steele as a "one eyed town" but a favourite spot with the Newfoundlanders. Set in the midst of the Cheviott Hills Stobs Camp had more sunshine than Edinburgh and air that was remarkably fresh and invigorating. The four months spent there, as Nicholson points out, were remembered as "the happiest of any similar period during the war."[32] At nearly 1300 all ranks the regiment was now at full strength and from all accounts remarkably fit. Days were filled with skirmishing, drills and physical exercise but there was also time for recreation – for sports meets and soccer matches with other battalions. Two events of note occurred at Stobs Camp: the presentation of a King's Colour, donated by the colony's Daughters of the Empire, and the recording on film of some of the regiment's activities.[33]

With the arrival of "E" company Steele took up his first appointment as a company officer. Before he left Stobs Camp for Aldershot he was shifted to "D" Company, his brother Jim's company, a move that pleased him a great deal as the two were especially close although often separated by the conventions of rank and military discipline. As a young subaltern Steele was obliged to take on some of the least popular duties of the commissioned junior rank including that of mess secretary-treasurer, a position he clearly disliked but was unable to escape. His responsibilities required him to manage the affairs of the officers' mess, which included canteen and accounts, certainly not onerous but, as he quickly discovered, time-consuming.

By the time the regiment left Stobs Camp in early August the war was a year old. Steele's entries show an eagerness for action even though he correctly anticipates that life at the front will be dangerous and may involve loss of life or wounding. As rumours abound that the regiment will be moving shortly the word comes down that it may be the Dardanelles. Given the fact that the Germans were apparently prepared to fight a long war, Steele sees action in Turkey as offering a shorter route to Berlin.

EXTRACT FROM LETTER DATED 17 MAY 1915
Since 'E' Company came over, I have been attached to it. Its officers are:-

Officer-Commanding Company — Capt. O'Brien
Second in Command — Lieut. Ledingham[34]
Lieuts. in Charge of Platoons — Allerdyce, Pippy,[35] Robertson[36] and Self
I think we shall make a very good company out of them.

We left the Castle last Tuesday, May 11th, and are now at Stobs, fifty miles from Edinburgh. We have fine camping ground and shall now be able to put in some good work. The day we moved, it rained all day and all that night, and, of course, it made us all very uncomfortable. It caused it to look very much like Salisbury Plains, but with only a hundredth part of the mud. It has now been fine for several days, but the nights are terribly cold. I have to-day hired a camp bed-stead. It will be up to-morrow. Most of the officers who were able to get one, have one. It is very uncomfortable and unrestful laying on the hard wooden floors of our tents with only a blanket under us.

There is a German Concentration Camp about a half mile away. I have seen it, and they say there are about 2000 German prisoners there. There are rows and rows of wooden shacks and they look as if the prisoners ought to be very comfortable. I went up with Eric Ayre.[37] I also went for a two hours walk with him this evening.

Our nearest town is Hawick (pronounced Hoyk) which is about three miles away. It is a one-eyed sort of place of about 18,000 people.

I am Secretary Treasurer for the Officers Mess. The appointment was made the week before we left the Castle at a Mess meeting, held when I was absent. I was at the time Orderly Officer for 'E' Company at Shrubhill Mills and therefore could not get there. I did not hear of it for two or three days. Someone proposed me and it was seconded and carried. The Colonel then said he thought that would be a satisfactory appointment and then asked for me, but when told that I was down at Shrubhill said "Oh! it will be a pleasant surprise for him then."

I have tried and tried to get out of it but cannot, for I did not want it, so I suppose I must stick it for a while.

EXTRACT FROM LETTER DATED 30 MAY 1915.
I have heard of Gordon Boone's [NI] death. I was speaking to him on Salisbury Plains last November. I also saw Jensen (who is wounded), Bert,[38] Snow [NI] and many others, when there.

When we go to the front, it will not be one Newfoundlander today, and one to-morrow &c., but suddenly you may hear of a whole Company being wiped out; for the above were perhaps one in this Company and one in that and so on.

At Xmas I had a Post Card from Mr. Currie,[39] President of the M.C.L.I. [Methodist College Literary Institute], and I replied, telling him that I would be unable to fill my position of Secretary. I also mentioned that the assistant Secretary was with us, viz. Colour Sergeant [George] Taylor.[40]

We have been here only nineteen days and it is rumoured that we are going to be again moved. Here, we have quite a nice Camping Location and are quite happy.

EXTRACT FROM LETTER DATED 10 JUNE 1915
Last Sunday, I had a great surprise, having received a letter from Mr. R. Dalton of Vancouver,[41] who is at present in London. He said he

Newfoundland Regiment presented with its colours, 10 June 1915

had just seen my name in the "Canadian" as having received a Commission.

I believe we may move now at any time, and I have heard that it will be probably to Bedford – 60 miles from London. Bedford is principally an equipping station, it is said, so that will mean that our time for going to the front is not far off.

This afternoon we are going to be presented with our Newfoundland Colours, by Lady MacGregor.[42] Jack Fox is going to receive them, he being the Premier's stepson.

It is grand out in the country like this; we are all well and happy.

Am very busy with Mess work, finalizing May account.

Postcard of the 1st Newfoundland
Regiment with the Regimental Crest

Regimental flags of the 1st
Newfoundland Regiment

EXTRACT FROM LETTER DATED 20 JUNE 1915

We are very busy now, for from the time we get up in the morning,
we are continually at it all day until evening Parade finishes, which is
generally about 4.45 p.m. Then at 5 p.m. there is a Special Officers
class for Musketry Instructions, Bayonet Exercises, Sword Exercises
and Physical Exercise which lasts for over an hour. We then have to
get ready for dinner which used to be at 6.30 p.m. but has lately been
changed to 7 o'clock.

EXTRACT FROM LETTER DATED 22 JUNE 1915

I saw in a paper to-day that the Germans claim that they have everything so systemized that they are fully prepared to carry the war on
for an indefinite period.

One or two of the officers have been away for a week or ten days
training, namely, Herbert Rendell,[43] Frank Knight and Jack Fox
(when he was Sergeant), also Bert Butler who was away for a signal-
ing course – he is in charge of the Signallers.

I expect that in about a month and a half, or thereabouts, we shall probably be going to the Dardanelles, and make for Berlin from that quarter.

EXTRACT FROM LETTER DATED 26 JUNE 1915

Re our Washings: As a rule, the Company Sergeant Majors arrange for the washing of those in their Companies. Jim gets his done by a woman who lives quite near our Camp (200 yards away) and she does any needed repairs. All the officers, or most of them, send their things to the Grocery Canteen.

In the photo you sent, I see mother is "Knitting socks for the soldiers." Neither Jim nor I are in need of anything at present, but you need not give away *all* the socks you knit.

The Fifth Contingent – F Company – is expected at any time. The tents for them were erected this morning.

Two Newfoundlanders visited us yesterday – Victor Gordon [NI], who is a Lieutenant in the Gordon Highlanders, and one of the Templeman's [NI] from Bonavista, who is a Lieutenant in a Royal Army Medical Corps. He has recently got his doctor's degree from McGill University.

EXTRACT FROM LETTER DATED 5 JULY 1915

Our company (E) has just started its Musketry Practices, and will be at it for about another ten days. We have to get up at 4.30 a.m. and after having Breakfast, march about three miles to the Rifle Range. We take our Lunch and Tea with us and do not return to Camp until about 8 o'clock p.m. The weather has been very good indeed and just now it is really glorious. I think Scotland a beautiful country both as regards its general weather and magnificient scenery.

Although F. Company was understood to have left St. John's on June 18th and reported to have arrived in Liverpool about June 27th, they have not yet arrived here. We were fully prepared for them on Saturday, June 26th, and the Adjutant said they would reach camp sometime Monday, since then we have heard nothing of them. There was a rumor that they were quarantined at Liverpool for Smallpox. Perhaps there is some truth in the rumor for it looks as if they have been in Liverpool now for over a week.

Victor[44] wants to know, when 'A,' 'B,' 'C,' & 'D' Companies go to the front, if I shall be going too. Well if I am not told to go, I am going to ask to go, for I consider it only right that all those who came over last October with the First Contingent should have the privilege of going to the front before those who volunteered and came over afterwards.

Shortly after I received my Commission, Sergeant [Walter M.] Greene[45] was made Provost Sergeant. He was originally a St. John's Policeman.

EXTRACT FROM LETTER DATED 19 JULY 1915

We are still at Stobs, but I do not think we shall be here much longer. There has been rumours lately of our moving, and last week the Colonel and the Quartermaster went to Ayr, which is, I believe, about half an hour's journey from Glasgow. Some seem to think that the whole six Companies are going there, and others say that only 'E' and 'F' Companies are going there and that A B C & D Companies are going South in preparation for going to the front. If the latter be the case, I am going to ask to be transferred to one of those companies. Of course, we may be transferred back before then, for there are three of us, namely, Capt. O'Brien (one of the best Captains in the Battalion), Lieut. Ledingham (the senior Lieutenant of the Battalion), and myself.

Bert Smith has now got a Commission. I saw him yesterday, having gone down to Hawick the latter part of the afternoon to see him, he was up from Edinburgh for the week-end. I was with him for about an hour. You will remember, I saw him last October on Salisbury Plains (he was a Private then) and some time afterwards he was taken sick and went to hospital with pneumonia. Whilst he was there, his Battalion the 72d Seaforth Highlanders (from Vancouver) went to the front, so when he came out, he was attached to the 17th Battalion and was given a job in the Orderly Room as Clerk where he remained until about a month ago when he received a Commission. He is now with an Officers' Training Corps in Edinburgh, but not in the Castle. Robin Smith [NI] is at present in Hospital in Hawick with a broken ankle.

It is now 4.30 p.m. and still raining and has been all afternoon. I have just come from a Lecture, which has been on all afternoon to Officers & N.C.O.'s, on Outposts, Sentries, Sentry Groups, Picquets, Supports, Reserves &c. The Lecture was given by Major Drew, our Second-in-Command.[46] During the past week we have had two new officers appointed to our Staff, – Major Drew and Major Whitaker,[47] – our Senior and Junior Majors. The former is about 55 years of age and the latter about 45.

EXTRACT FROM LETTER DATED 1 AUGUST 1915 (STOBS CAMP)

Well I expect this will be the last time I shall write from Stobs, for A, B. C & D Companies go South to Aldershot to-morrow, Monday evening, and the reserves go to Ayr, near Glasgow. As I think I told you before, if I were not transferred to one of the Companies going to the front, but was left to go with the Reserves at Ayr, I was going to go to the Colonel and ask to be placed with one of the Companies of the Battalion. When last night's orders came out, I had the pleasure of seeing my transfer from E Company to that of D (Jim's Company), so am happy on that matter now.

On Friday night, the Four senior Lieutenants, namely, Ledingham, Raley, Ayre and Rowsell,[48] were made Captains. They have been holding the position, viz. Second-in-Command of Companies for the past three months but had not the Rank that goes with it until now. All promotions in the Commissioned ranks are made according to Seniority, except, of course, for anything special, like on the battlefield.

Major Whitaker is going to our Depot, Ayr, as Officer Commanding, Capt. Ledingham as acting Quartermaster, and Lieutenant [S] Robertson as acting-Adjutant. I would rather go to the front than to stay back in any position. Capt. O'Brien is also remaining behind and is very sore over it. – I think we shall be going to the front in less than three weeks.

CHAPTER THREE

Aldershot to Egypt
(3 August 1915 – 14 September 1915)

The Newfoundlanders were sorry to leave Stobs Camp and Scotland. Many close relationships had developed between Newfoundlanders and Scots, as evidenced by the warm send-off the regiment received from the people of Hawick. Years later, many looked back at the time in Scotland with great fondness. Perhaps it was the people, or the rugged countryside that reminded so many of the Newfoundlanders of home. For the regiment the time in Scotland was important because it was there that it developed a strong sense of identity, with bonds formed among the soldiers that would carry them through the hard months that followed. The regiment was now up to strength and its future assured, as least for the moment. It left for Aldershot in the best of spirits and ready for the final stage of training before it entered the line at Gallipoli.[1]

Aldershot was well known as Britain's premier training camp and the final stop before posting to the front. The regiment was housed in the historic Badajoz Barracks, Wellington Lines, connected with so many of Britain's famous regiments over the years. Quarters were comfortable with rooms large enough to hold a platoon of three NCOs [Non Commissioned Officers] and twenty-eight men. The time spent in Aldershot was short, a mere seventeen days, but sufficient for the final toughening of men eager and ready for combat. While the regiment was at Aldershot Lord Kitchener, Britain's greatest soldier and now War Minister, appeared to inspect the 22nd Division, to which the Newfoundlanders were temporarily attached. The Newfoundland Regiment was the only one he addressed: "I am sending you to the Dardanelles shortly," he announced, "so be prepared, for when the order comes it will come sharply.[2]" At long last the regiment was being sent into action, creating a sudden burst of excitement in the ranks. As the men of A and B companies had completed their year of service – the initial requirement of enlistment – they were asked to sign on for the duration of the war, standard practice in all the volunteer units raised in Britain and abroad at the outbreak of the conflict. Most signed on.[3]

At Aldershot the regiment received its tropical kit, described by Mayo Lind as "very thin cotton stuff, khaki colour, and khaki helmets." His impression was that the Newfoundlanders looked "some stylish" and that their helmets would create a rush of "Helmet-shape" designs in women's wear the following year.[4] As Steele recounts, it was also at Aldershot that the regiment received its immunization shots, a novelty for most. One soldier, as reported by Nicholson, was certain that he was now fully prepared to meet the enemy: "if all the Turks start to decay at the same time, they won't affect me."[5] No one, of course, had any idea of what to expect in the far-off Gallipoli Peninsula, but the regiment was better prepared at this point for battle than it had been a year previously. As Lieutenant-Colonel W.H. Franklin, a former commanding officer, later recalled, having watched the regiment return from a morning's hard training at Aldershot: "They have become very soldierly and look ever so strong. The officers have improved in looks and bearing. Altogether they are a regiment anyone should be proud to command. They will make a name for themselves."[6]

The battalion embarked at Devonport on Friday, 20 August 1915 joining 1,000 of the battle-hardened Royal Warwickshire Regiment on board HM Transport *Megantic* of the White Star line. The first phase of the trip to Malta was without incident and pleasant. The *Megantic* was much larger than the *Florizel* and far more comfortable, as Steele notes, and friendships quickly formed between the men of the two units. The first night, two destroyers escorted the troopship across the Channel. Passing the Straits of Gibraltar the ship began a zig-zag course through the Mediterranean to avoid enemy submarines. When the ship docked in Valetta Harbour overnight, commissioned officers were permitted to leave the ship and Steele used the occasion to enjoy a meal and attend a theatre performance. Other ranks remained on board, entertained by thoughts of what shore life might be like and tormented by Maltese "bum boats" selling wares and young boys willing to dive for coins thrown from the ship.[7]

Two days later the *Megantic* anchored in the enormous but exposed harbour of Mudros on the arid, thinly populated Aegean island of Lemnos. Mudros was the base for Allied operations in the Eastern Mediterranean and a gathering point for troops destined for the Gallipoli theatre, some 80 kilometres distant. The Newfoundlanders expected to land at Mudros and, fully kitted out, were prepared to do so.[8] Most, like Steele, interpreted the decision to proceed to Alexandria as the result of a blunder somewhere up the line. In fact, as Nicholson points out, the last minute change in plans resulted from more pressing concerns and the need to bring the Warwickshire Regiment to Khartoum quickly as a reinforcement for General Sir John Maxwell's Egyptian defence force. The Newfoundlanders were disappointed by news of the change, as Steele mentions, but it was only a brief reprieve from the conflict as the troopship headed south to Alexandria and a period of acclimatization for the Newfoundlanders in Egypt.[9]

Egypt was a British protectorate at the time, having been occupied since 1882. The Newfoundland Regiment was sent first to Cairo, where it was housed in Abbassiah Barracks,[10] constructed during Lord Kitchener's incumbency as Sirdar or commander-in-chief of the Egyptian Army in the early 1890s.[11] Their next-door neighbours were the Australians, who had already established a reputation as fierce fighters from their time in the Dardanelles. The two groups of "Colonials" got along famously, forming warm relationships that would continue while they served together in Gallipoli.[12] For two weeks, as the regiment got acclimatized, it remained at Cairo and then proceeded to Alexandria awaiting embarkation for Gallipoli.[13]

The Newfoundlanders had mixed feelings about Egypt. The heat was terrible and the flies and filth exasperating – yet the country was appealing, with its long and rich history and its strange customs. Visits to the pyramids and other historic sites, many with biblical connotations, brought to life stories about ancient Egyptian civilization and Moses and the captive Israelites learned in school or in church. Lind's account is alive with details about the novelty of the country's rich heritage, its peculiar customs, and his personal encounter with a remarkably different country.[14] Steele, on the other hand, was unimpressed with the pyramids, his imagination far exceeding the reality of what he saw in the Egyptian desert. His account is short and remarkably reticent about his impressions of the country and its people. Perhaps the lack of detail is the result of his having been exceptionally busy with preparations for the move to Gallipoli and a theatre of war. His brief entries reveal restlessness and also apprehension, especially as the time came to embark for Suvla Bay. He was clearly eager to get underway and into the fighting, and yet he realized that the risks were great and that other units had suffered great losses on arrival without even firing a shot. It was a feeling echoed by Mayo Lind in his somewhat restrained report to the *Daily News*: "May we do our part and bring honour to Terra Nova; it will not be all fun, for my opinion is we have had the best of it so far, and now comes the sterner side."[15]

EXTRACT FROM LETTER DATED 3 AUGUST 1915
Badajoz Barracks, Aldershot.
We arrived here at 7 o'clock this morning, having left Stobs 7 o'clock last evening. It is believed we shall be here for only three weeks and will then be off for "somewhere" at the front.

EXTRACT FROM LETTER DATED 9 AUGUST 1915
Badajoz Barracks, Aldershot.
We see Aeroplanes every day now, sometimes four or five at a time.[16] This afternoon we marched two or three miles in the country and did some attack work.

Since being moved to D Company, I have had charge of No. 15 Platoon. Cyril Duley[17] is my Platoon Sergeant. Jim is in No. 14 Platoon.

Lieutenants Keegan,[18] Rendell and Rowsell have gone home to train G Company.

THURSDAY – AUG. 19TH/15
Left Badajoz Barracks, Aldershot at 9.45 p.m. and marched to the Railway Station. Left Aldershot at 11.20 p.m. and arrived at Devonport at 9.30 a.m. on Friday. Weather: sunny & warm.

FRIDAY – AUG. 20/15
Arrived at Devonport at 9.30 a.m. and after Roll Call went on board the Troopship (White Star Line) "*Megantic.*"[19] Spent all day in getting men fixed off and getting ammunition, &c. on board.

Left Devonport at 6.40 p.m. accompanied by two destroyers, one on either flank as protectors. These destroyers accompanied us until 5 o'c Saturday morning. Weather: sunny & warm.

EXTRACT FROM LETTER DATED 20 AUGUST 1916.
On board train, (6 a.m.)
From 6 a.m. to 5 p.m. (men's tea time), the officers have been constantly busy, getting their men fitted out with dozens of various necessaries for the front. They have been issued with new boots, (better quality than previously), new uniforms (Khaki Drill, light in weight and colour) and Helmets. Their stocks of the following

were also completed:- Socks, Shirts, Underclothing, Housewives, Canteens, Water bottles, – in fact, twenty-five or thirty other items. Sometimes we were compelled to work after tea. Then what spare time I had, I endeavoured to get the Mess Books done up. However, I finished and got clear of them yesterday at lunch time – and am not sorry.

We 'fell in' last night at 9 o'clock, marched to the Station and left Aldershot at 11.20 p.m.; that is, C and D Companies, for A and B left an hour before.

The new Uniforms, Boots and Helmets are not being worn just yet, but have been all packed up in boxes. The officers have had to get a lot of new equipment also, but have had to pay for it themselves.

Our new uniforms cost us about £4-15/-, Helmets 21/-, Boots 30/- to 40/-, Binoculars and Compass, 5, and £3-10 respectively, I think.

Our next address will be;

First Newfoundland Regiment,
C/O General Post Office,
58 Victoria Street,
London,
England
British Mediterranean Expeditionary Force.

EXTRACT FROM LETTER DATED 20 AUGUST 1915

On board S.S. "Megantic"(5 p.m)

This is just a final word before we go; as you see from above address, we are on board the White Star Liner *"Megantic"* (15,000 tons) and are all ready to go, and may do so any minute, but expect she will leave about 6 p.m. We are having a Warship go with us and two destroyers for a period of twelve hours. I think we shall be about a fortnight on the water, but will be a thousand times more comfortable than on the *"Florizel."* The *"Arabic,"* which was torpedoed yesterday, is of the same line.

SATURDAY – AUG. 21ST/15

This morning was spent in preparing lists of Deficiencies in men's clothing and equipment.

This afternoon was occupied in apportioning and allotting men to Boats and Rafts, should it be necessary to use same at any time during voyage.

At 6 p.m. the "alarm" was sounded for a "Boat Drill" practice test.

There was on board, besides our Battalion of 1033 men, a battalion of Warwickshires.

Weather: Sunny & warm.

SUNDAY – AUG. 22ND/15
Church Parade was held this morning conducted by the Chaplain of
the Royal Warwickshire Regt. This Regiment was on board with us
and was of about the same strength.[20]
    Weather: Fine & very warm.

MONDAY – AUG. 23RD/15
Boat Drill Parade was held at 10.30 a.m. Nothing else doing today.
Weather: fine & very warm.

TUESDAY – AUG. 24TH/15
Same as Monday.

WEDNESDAY – AUG. 25TH/15
Nothing doing today. Saw several warships at night. Weather: fine &
warm with nice cooling breeze all day.

THURSDAY – AUG. 26TH/15
Arrived off Malta at 11.30 a.m. and had to stand by & await orders,
but keep moving. At noon commenced to go into Valletta (Capital(?)
of Malta), and got to our berth at 1 p.m.
    At 4 p.m. officers were allowed to go on shore until 6 p.m. I went
ashore with Gerald Harvey[21] and meeting [Bertram] Butler and
Windeler[22] we hired a carriage and went for an hour and a half's
ride. Saw the Governor's house and walked thro' the Marine Gar-
dens. Malta is of very Eastern style.
    Went aboard at 6 p.m. and found we could remain until 11 p.m. so
went ashore again. Had dinner and then went to a small Theatre and
was on board at 11 p.m.
    Weather: fine & warm.

FRIDAY – AUG.27TH/15
Left Malta for Island of Lemnos at 7 a.m.
    Weather: Sunny & warm.

SATURDAY – AUG. 28TH/15
Passed several steamers and warships. Nothing unusual. Weather:
fine & sunny with cool wind.

SUNDAY – AUG. 29TH/15
Arrived at Mudros, – Island of Lemnos, – at 7.30 a.m. and anchored.
Moved in one of the many small harbours at 10.30 a.m. and

Newfoundland Regiment on board a troop ship off Mudros, 1915

anchored again to await orders. Having received orders we left for
Alexandria at 6.30 p.m. – Through some bungling at Malta we had
been sent to Lemnos by mistake.

Weather: very warm and sunny.

EXTRACT FROM LETTER DATED 29 AUGUST 1915
On board S. S. "Megantic."
I have not very much to say, simply because we must not. However,
the main thing is, we are both well and happy and are enjoying ourselves.

The weather has been glorious all the time, and, on the whole, has
not yet been too warm.

I could write quite an interesting letter, but must wait until we
meet when we shall have many a pleasant hour, recounting our experiences and impressions.

I hope all at home are as well and happy as we. Do not worry
about us, we shall be O.K, and shall take good care of ourselves as
far as possible.

MONDAY – AUG. 30TH/15
Had another excellent day's passage. Nothing unusual occurred.

Weather – As usual – Fine and sunny.

TUESDAY – AUG. 31ST/15
Arrived at Alexandria at 2 p.m. At dinner time – 7 p.m. – Three officers per company were permitted by the Colonel to go ashore. More
than this number went, but I had to remain behind, having been
warned to go on guard at 7 p.m. & to remain on until 7 or 8 tomorrow morning.
Weather: Fine and extremely hot.

WEDNESDAY – SEPT. 1ST/15
I came off guard about 10.30 a.m. After a wash, etc., asked the
Colonel's permission to go ashore. He told me I could go for a short
while – left ship about 11.30 a.m. & returned 3 p.m. – Had a guide
who took me through the native quarter first, then through the European. Very dirty city and very dirty people and unattractive surroundings. Went on shore again at 8 p.m. and remained until 10.30 p.m.
Impressions unimproved. Boarded train for Cairo at 12 midnight.
Weather: – very fine and very hot.

THURSDAY – SEPT. 2ND/15
Left Alexandria for Cairo at 1 a.m. and arrived at our destination,
Abbassia[h] Barracks, Abbassia[h] (just outside Cairo) at 8 o'c. Spent
day in getting settled.
Weather: Exceedingly hot & fine.

FRIDAY – SEPT. 3RD/15
Paraded at 6 a.m. and had a few miles march in Full Marching
Order, returning about 7.30 a.m. 4.30 p.m. Rifle Inspection.
Weather: As Thursday. Plenty of flies.

SATURDAY – SEPT. 4TH/15
No parades today, weather too hot.
Weather: as above. Flies a perfect nuisance.

SUNDAY – SEPT. 5TH/15
Paraded at 4.45 a.m. and with Fatigue Parties proceeded to Polygon
Camp – 2 miles away and proceeded to erect tents, and about 6 p.m.
moved the Battalion into Camp.
Weather: As hot as ever.

MONDAY – SEPT. 6TH/15
Paraded 6 a.m. and did some Field work for an hour an a half. Dur-
ing afternoon the Officers moved into Camp.
Weather: Hotter than ever.

TUESDAY – SEPT. 7TH/15
Paraded at 6 a.m. – 2 hours Field Work.
  Paraded at 4.30 p.m. – 1 hour Field Work.
  Weather: Very hot and fine.

EXTRACTS FROM LETTER DATED 7 SEPTEMBER 1915
Now for a few details:- As you already know, we left Devonport on
August 20th. On August 26th we arrived safely at Malta, having
taken a very zig-zag and erratic course to avoid submarine dangers,
still one or two were reported to have been seen. The *"Megantic"*
was chased four times on her previous trip. At Malta we received
orders to proceed to Mudros, on the Island of Lemnos, which is
about thirty miles from the Dardanelles; sometimes the guns can be
heard there. On arrival there, we were told we should not have come
to Lemnos and that someone had bungled, so we had to go to
Alexandria. Here we had to wait off shore for orders, and thus had
once more to face the submarine dangers. After waiting off for half
an hour we then moved to the wharf at 2 p.m. August 31st. We
remained at the wharf until 12.30 a.m. September 2nd when we disembarked and proceeded by train to Cairo, arriving there at 8 a.m.,
and went into Barracks just on the outskirts of Cairo.

  Our present Camp is, of course, in the Desert, for there is Sand
and Desert everywhere here. There are many troops all around this
District in Camps and Barracks. There are dozens of magnificent Barracks around here which have been built during the past three or four
years. Lord Kitchener certainly knew what he was doing, and evidently foresaw this war, when he had them built. Ten minutes walk
from here takes us to the street cars and fifteen minutes more takes
us right into Cairo.

  So much for where we are and how we reached here. Now to go to
our sea trip again. First of all, it was a glorious trip from start to finish; we had bright, warm and sunny days ever since leaving England,
all the sea was as smooth as a pond all the way. Inside the steamer
one could hardly believe that he was on the water, for there was not
the least movement in the steamer. We were all sorry when the trip
came to an end, despite the dangers from submarines – for we all
thought "the game was worth the candle." Of course, we had boat
drill everyday; there were nearly 2,500 in all on board, for whom
there were boats and rafts. Six rafts in addition to boats were allotted
to our Regiment and I was placed in charge of one of these with an
allotment of 25 men. (These rafts were only about 10 feet by 6 feet.)

  We were all very disappointed at not going to the Dardanelles,
and to think that we were within sound of the guns and then taken

away again; each man had even been issued with his 120 rounds of ammunition.

Wherever one goes among British soldiers, one hears the query: "Are we down-hearted?" to which the reply comes from everyone in the long drawn out word "No!" Well, when we were at Lemnos, there were hundreds of warships and other troopships in the various harbours; our men used to shout out to the other troops this query and the reply immediately came "No!" – when they had finished our boys would shout, "Yes, we are, we're not going to the Dard-anelles." However it is the general belief that we shall not be here for long.

The weather here is extremely hot, being nearly 100 degrees in the shade in the middle of the day, and often goes past 115 degrees. It is far too hot in the daytime to have any Parades and there is an Egypt-ian Standing Order regarding the Military, that no work of any sort is to be done between the hours of 9 a.m. and 4.30 p.m. There are hundreds of various "Egyptian Standing Orders" and they are really Military Regulations for Egypt.

During the day time we can do nothing but lay off in our tents, which are double ones with a space of about a foot between each to protect us from the scorching rays of the sun. Even in this laying off, there is no pleasure, for one perspires then too, and then there are the flies which are very, very annoying. They are only the ordinary house fly we get at home, in appearance and do not bite or sting, but the skin seems very sensitive, perhaps owing to the heat.

WEDNESDAY – SEPT. 8/15
Paraded at 6.30 a.m. – 1 ½ hrs Field Training, & 4.30 p.m. Ditto.
　　Weather: very hot & fine.

THURSDAY – SEPT. 9TH/15
Paraded at 6.30 a.m. for an hour and a half and had no more for the day.
　　In the afternoon arranged a party of 12 and went to the Pyramids. Were all disappointed, for having read and heard so much of them, we expected more than the reality.
　　Weather: hot but not quite so hot as to-date. perhaps getting a lit-tle used to it.

FRIDAY – SEPT. 10TH/15
Paraded at 6.30 a.m. for a couple of hours.
　　Weather: As Thursday.

SATURDAY – SEPT. 11TH/15
No Parades at all today.
Weather: A little cooler and windy, but warm enough to cause one
to perspire even when lying off in Camp. Flies worse than ever.

SUNDAY – SEPT. 12TH/15
Church Parade at 9.30 a.m. Catholics marched a mile to Heliopolis.
Protestants, in conjunction with our neighbours (London Regt.) had
service on Desert near Camp.
Weather: very hot again with plenty of flies.

MONDAY – SEPT. 13TH/15
Paraded at 6 a.m. for Inspection by General & his Staff. – lasted two
hours.
All the men's equipment, and what not required of Officers, taken
by the Transport to Abbassia[h] Siding ready for shipment to Alexandria as the Battalion is under orders to embark from there tomorrow
for an unknown destination.
Weather: Not so oppressively hot.

CHAPTER FOUR

Suvla Bay
(19 September 1915 – 5 December 1915)

In the night they let us wander,
There was no one sent to meet,
Till we found some empty dugouts,
Where to rest our weary feet,

Airplanes dropped bombs upon us,
Shells went screeching overhead,
Shelter from the rain was asked for,
There is none the staff all said.

Then the order came to dress up,
To get picks and shovels from HQ
They brought back no bloody shovels,
And I fear the story is true,
For they raided and stole puddings,
Pinched the General's turkey too.
And for Christmas cheer they gave us,
Bulley beef and biscuits few,
No tobacco, rum or pudding,
Did we grumble – wouldn't you?

All our lousy shirts and jam tins,
Or the parapet did throw,
Did the staff complain about it,
No! they chucked it long ago.

Capt. Wilson says we are dirty,
Armstrong's views are just the same,
So we are never downhearted,
What is dirt compared to fame.

Song composed for the Dardanelles Companies[1]
(Tune: What a Friend We Have in Jesus)

The Gallipoli offensive was a complex affair that has been examined at great length elsewhere.[2] It was intended to relieve pressure
on the Western Front by driving Turkey out of the war and bolstering
Russia's efforts on the Eastern Front. The 1916 Royal Commission
concluded its report with the remarks: "from the outset the risks of
failure attending the enterprise outweighed its chances of success." In
the 259 days between the first landings in April 1915 and the evacuation of January 1916 a half million men had been sent to Gallipoli.
More than half were casualties. British forces alone, of which the Newfoundlanders were a part, suffered 205,000 casualties. The campaign
also destroyed several reputations, Kitchener, Hamilton and
Churchill's among them.[3] The naval operation – a botched attempt at
forcing the Straits – actually began much earlier, on 19 February. By 22
March it was already in trouble and ground forces, under the unfortunate General Sir Ian Hamilton, were shortly thereafter committed to a
difficult landing on the Gallipoli peninsula. On 25 April the first
British landings, by the 29th Division, occurred at X, W, and V Beaches at Cape Helles, on the southern tip of the peninsula. Those landings
got nowhere, meeting stiff Turkish resistance and facing an impossible
topography over which to fight. In August, Hamilton shifted the operations of the 29th Division to Suvla Bay, a point about 15 kilometres
farther up the peninsula. It became, as John Keegan has described it,
"the third shallow and static enclave maintained by the Allies on the
Gallipoli peninsula." By then the Turks had strengthened their defences
and were firmly entrenched in their positions on the high ground.[4]

The Regiment's baptism of fire came during the Gallipoli expedition.
Few had the remotest idea of what lay ahead – months of endless digging and trench warfare, chronic sickness and privations, bleak surroundings and harsh climate, and, above all, the constant danger.[5] The

Newfoundland Regiment was the only non-regular volunteer unit in the British 29th Division[6] (the "Incomparables"), which presented a special challenge to the inexperienced colonials. For three months the Newfoundlanders endured the dangerous and inhospitable confines of Suvla Bay, a barren bit of unprotected shoreline exposed to the lethal fire of distant Turkish batteries. The routine was deadly – four days in trenches, eight days in rest dugouts, and then back to the trenches. "This trench warfare is really very monotonous," Steele complained in his diary, "not comparable to 'Guerilla' warfare, as that of the Boer War, where one would get lots of excitement." The hard experience of front-line duty was costly: forty-three Newfoundlanders, including several officers, were killed. Suvla Bay left enduring memories of personal privation and homesickness. "The talk was much of home" John Gallishaw recalled, "a poignant longing for those dark, cool forests of pine where the caribou roa.m."[7] Prolonged delays in receiving letters and parcels from home kept spirits down.[8] In spite of the hardships, Suvla Bay was a valuable learning experience under battlefield conditions that served the Newfoundlanders well when they entered the European theatre.[9] When the Allies finally evacuated Suvla Bay and later Helles the Newfoundland Regiment performed the important and dangerous role of rearguard and were among the very last troops to leave the Gallipoli Peninsula, during the night of 8–9 January 1916.[10]

Steele's first encounter with the face of battle was mixed. He was clearly shaken by the death of Lt Taylor [NI] of the Royal Dublin Fusiliers, with whom he had shared a pleasant night watch. Taylor's death showed how unpredictable war was and how much they were all part of a terrible game of chance. And yet he felt energized by preparations for a Turkish attack – which failed to materialize – and was astounded by the might of the combined naval and artillery bombardment of enemy positions. He began to understand the risks of working in exposed positions, yet seemed to take a certain pride in his ability to avoid being hit, much of which was the result of either excessive bravado or just plain naiveté. When the regiment was ravaged by fever and diarrhoea he thought at first he had escaped it because he followed the "fresh air course." A week later he was laid low and took nearly five days to recover.

By November Steele was getting bored with the routine and the absence of a serious threat from the Turks. Far removed from the real action taking place in other sectors of the Front, the Newfoundland Regiment had to make do with occasional raids and sniper patrols. Its only "action" was against a Turkish position known as Caribou Hill, which won the regiment a Military Cross and two Distinguished Conduct Medals.[11] The terrifying storm on 26 November with accompanying flooding and bitterly cold temperatures was a never-to-be-forgotten expe-

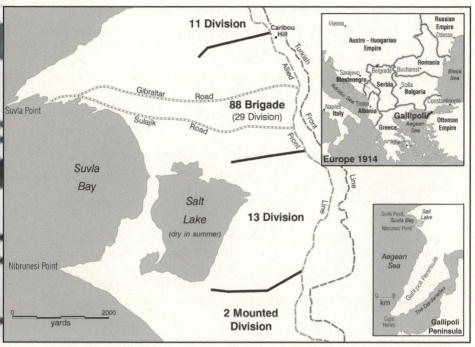

Suvla 1915, showing the opposing trenches in mid-December

rience." The flooding was so severe that it swept away stone parapets, animals and men, British and Turk alike. Trenches quickly filled with a raging torrent of water and ice, forcing soldiers on both sides to evacuate. Those who did not were drowned in the flood. For the next two days icy winds struck a further blow – soaked clothing simply froze on wearied soldiers. As one witness recounted, "Many were frostbitten; many lost their limbs; some, their reason."[12] An officer of the [4th] Worcesters[13] found some thirty men frozen to death on the firesteps of a trench. At Suvla there were over 12,000 cases of frostbite and exposure. A total of 280 died as a direct result of the storms.[14] Within the Newfoundland Regiment, upwards of 150 men ended up in hospital, most with frostbitten feet. It took days for fighting lines to be restored.[15] News of a possible evacuation two weeks later was therefore greeted with considerable relief.

TUESDAY – SEPT. 14TH/15
Reveille 4 a.m. to prepare for two Companies to leave at 6.30. Remaining two Companies left at 7.30 a.m. My Company was in the latter two and we left Abbassia Siding at 9.30 a.m. arriving at Alexandria about 3 p.m. My Platoon did not get on board steamer

until after 5 p.m. Our steamer was a very poor one – the "*Ausonia*"
a 4 yrs old purchase by the "Cunard Line."[16] Our Camp neighbors –
The London Regt. – (Reserves) – were also on board. We left
Alexandria at 6.30 p.m.
Weather: Fine and very hot.

WEDNESDAY – SEPT. 15TH/15
We understand we are going to Lemnos once more. Are making a
very zig-zag course.
Weather: Now much more pleasant and windy.

THURSDAY – SEPT. 16TH/15
Have had Boat Drill Parade but to tell the truth have far from
enough accommodation.
    Still making a very erratic course. Sometimes going West sometimes
East and sometimes *South* – though our destination lies North. Some
say we went South last night for a period of four or five hours – no
one understands the why or the wherefore of our erratic course.
Weather: Now quite normal but still fine.

FRIDAY – SEPT. 17TH/15
Passed one or two islands this morning about 5 or 6 a.m. had quite a
lot of rain with thunder & lightning, – first rain since leaving England almost a month ago.
Weather: Fine & warm after the rain storm.

SATURDAY – SEPT. 18TH/15
Arrived at Mudros Bay Island of Lemnos, at 11 a.m. Spent day in
making preparations for disembarking. Received first mail since leaving England about 8 [illegible]. My share was ten letters.
Weather: Fine and not too warm.

EXTRACT FROM LETTER DATED 18 SEPTEMBER, 1915
We are going to Suvla and shall land on the Beach. We shall then
probably march about a mile and a half to the Reserve Trenches,
and after a few days shall watch our chance to get up to the firing
line.
    We have had a very good time all along so far, but we all know
that the hardest part has now to come. The place where we are to
land is shelled all day long, and the last Division which went there
lost 1200 men and 36 officers the *first day*, and that, without having
fired a shot, nor seen a single Turk, so we have heard.

SUNDAY – SEPT. 19TH/15

The morning was spent in continuing disembarking preparations. Transferred to the small steamer *"Prince Abbas"* (about the size of *"Fiona"*) at 3 p.m. and left for Suvla Bay at 3.30 p.m. where we arrived at 9.30 p.m. After an hour or more landing by means of Lighters [large barges] was commenced. I was in the last load which left the ship about 12.30 midnight.

Weather: Fine & warm during the day cold at night.

MONDAY – SEPT. 20TH/15

Left the *"Prince Abbas"* on a lighter at 12.30 a.m. After reaching the landing stage, owing to the unhandiness of these lighters it took two hours to berth us.

Landed about 3 a.m. and after forming up were led by the Landing Officer to our "Dug-outs" for the night, amid clouds of sand – just like a Nfld snowstorm. – Our "Dug-outs" are simply holes dug into the ground with no covering. We rolled ourselves in our blankets and slept as well as we could, for it was bitterly cold. Got up at 6 a.m. having had 2 hrs. sleep. We were shelled by the Turks for an hour, from 8 to 9 a.m. from a distance of 7 or 8 miles. We had some 14 casualties, including the Adjt.-Capt.[Walter Frederick] Rendell, who was during the day sent back to Mudros. Fortunately, no one was killed. Harvey, Knight and Self had narrow escape when having breakfast – a shell burst not 10 yds from us – shrapnel falling all around us.

During the day we moved behind a hill to be sheltered from the enemies guns, as we were on the side of a hill facing them. At night A. Coy. went to the Trenches.

Weather: Fine & warm during day but very cold at night.

TUESDAY – SEPT. 21ST/15

Spent day in preparing "Dug-outs" for stores etc. During the after-noon Capt. March & I went down and had a bathe in the Bay. In the evening the Turks sent a few shells over but did us no harm.

B. Coy. went to the Trenches tonight.

Weather: Same as above.

WEDNESDAY – SEPT. 22ND/15

Spent today in building Terraces to make shelters for the men as a greater protection against shell-fire. Scores of shells & shrapnel were sent over our way at 3 p.m. for over an hour, quite a number of which dropped among our "Dug-outs" (Harvey, Knight and I are in

together). We had one man killed (McWhorter of 14 Platoon D. Co.)[17] and 5 wounded. At 6 p.m. -Harvey & I proceeded to Ordnance Depot on the beach to attend to the storing of the Battalion's Kit-bags, etc. We returned about 11 p.m. and then had to spend the remainder of the night superintending the building of the Terraces being built. About 3 o'c. (on 23rd) we had to stop for half an hour as the Turks sent a shell over us. C. Coy. went to the Trenches to-night.
Weather: Same as above.

THURSDAY – SEPT. 23RD/15
We did not do very much work today as we were to go into the Firing Line Trenches tonight. About 8 o'c. we (D. Coy.) started for the trenches & arrived at the communication trench about 9 o'c. which was half an hour's march from the firing line to which we proceeded almost immediately. I went on watch in the trenches with Lieut. Taylor of the Royal Dublin Fusiliers (as officers of the watch) from 1.30 to 5.30. We both went out to a new trench we were building 50 yds in front to see the listening patrol about 3.30 a.m. (24th), but could not locate them, so came in again. About 4 o'c. Taylor went out again, he saying I need not bother coming, as we had been talking to another officer who was still with us. He had not gone five mins. when he was shot and died before 5 o'c., being buried shortly afterwards. It was only a matter of luck that I was not with him.
Weather: Fine & warm, but night very cold indeed. Flies were terrible, particularly at meal times.

FRIDAY – SEPT. 24TH/15
Absolutely nothing doing during the day, except occasional sniping, but during the night sniping was rampant.
Weather: Same as Thursday – also flies.

SATURDAY – SEPT. 25TH/15
Same as Friday. (We had one man wounded today.)
Weather: Same as Friday.
At 6.30 p.m. we (D. Co.) marched out of the Firing Line and went to some new "Dug-Outs" near the beach where the other three Coys. were – to have a day's rest.

SUNDAY – SEPT. 26TH/15
Spent today resting & writing letters – I wrote one home only, the sun was too hot and the flies too numerous and annoying to write more.
At 7.30 p.m. we marched off to a new part of the firing line, which we shared with the 2nd Hants [Hampshire Regiment]. The communi-

cation trench is the finest trench I have yet seen, being about 8 to 10
feet high, and about three miles long. – They represent a vast amount
of work to have been done in so short a time for the landing here is
only about 7 weeks old. Our times of watch (officers) were different
here for we had one hour on Duty & one hour off. This was, to my
mind, worse than being on duty all the time, for I would just be
beginning to doze when my time to go on again arrived. A few
snipers crept almost up to our trench, in a gully and threw 3 bombs
on to our parapet but did no harm. We had to get one of our
machine guns on to them, and also threw a Trench Mortar. Snipers
were, as usual, very busy all night.

Weather: Fine & warm and night not so cold as usual.

MONDAY – SEPT. 27TH/15
Heard that Lt. H. Rendell was wounded last night.

Did not have very much to do today. Inspected Rifles a couple of
times and had the trenches cleaned up ready for leaving tonight. A
few shells came over us this evening & burst dangerously near but
did no harm. In the day time we have one man in six doing Sentry
– he stands on the firing step and does what sniping he can – and
at night we double sentries having one in three. At 6.30 p.m. we
were ready to leave, but just then the Turks & Australians got
"wind" up (that is, each side got up rapid fire and each thought
the other was attacking) about 2 miles on our right flank. Not only
were the rifles snapping like bunches of Chinese fire crackers, but
Field Guns, Machine Guns & Naval Guns from the four or five
warships in the Bay, also took very active part. This, as is usual,
worked right along the firing line trenches until it reached us. We
had every man in our firing line (D. Coy.) on the firing step, with
Bayonets fixed and Rifle Magazines fully loaded in case the Turks
did decide to make an attack, but each Platoon Commander gave
his men strict orders to on no account fire without orders. The
result was that it did not go further along the line than our position. It lasted for about 2 hours and whilst it was on was certainly
very exciting and gave one an idea of what a real attack would be
like.

What with Rifles, Bombs & Hand Grenades, Trench Mortars,
Machine Guns, Field Guns & Flares which were constantly being
sent up by both sides to light up the field of activity, the scene was
very impressive indeed. The result was that we were 2 ½ hours late in
leaving the Trenches and did not leave until 9 p.m., reaching our
Dug-Outs about 10 p.m.

Weather: Fine and warm.

TUESDAY – SEPT. 28TH/15
At 6.30 a.m. went to one of the Communication Trenches with a
fatigue party of 50 men, to make some improvements – returned
shortly before dinner time. We were relieved by another party and so
on throughout the day and night.
Weather: Fine & warm.

WEDNESDAY – SEPT. 29TH/15
Same as Tuesday.
Weather: Fine & warm.

THURSDAY – SEPT. 30TH/15
Our Company did fatigue work until about 3 p.m. when we stopped
to prepare for going into the Trenches again this evening.
At 7.30 p.m. we left for the Support Trenches, where we arrived
about 9 p.m. A & B Coys. were in the Firing Line with C & D in the
Supports. We were told we were to maintain the Firing & Support
Lines for 3 weeks, C & D changing places with A & B every four
days. The officers of D. Coy. then arranged their watches. Mine was
from 3.30 to 6.30 a.m.
Weather: Fine & warm.

FRIDAY – OCT. 1ST/15
Not much doing today – Inspected Rifles. A. Coy. lost one man, Sam.
Lodge.[18] At 3 p.m. we were surprised to receive orders to prepare to
leave for our Dug-outs at 7.30 p.m., and no one knows the reason
why, but believe it to be on account of something affecting our
Brigade – the 88th. The Regiment which was to relieve us was very
late in arriving, with the result that we did not leave until 10.30 p.m.
arriving at our Dug-outs about midnight. On arrival our Coy. was
asked to supply a fatigue party at 7.30 next morning. I was asked to
take charge of the detail of 50. We had to make firing recesses in
about 400 yds. of the Communication Trench, so did not have much
time for sleep tonight.
Weather: Fine & warm.

SATURDAY – OCT. 2ND/15
At 6.30 a.m. we were unable to start for that work as the cooks were
late with breakfast. I then went to the Brigade Major and told him we
could not start until 9 o'c. We left at 9 o'c. and worked until 3 p.m.
when we finished and returned. In the evening our Coy. was again
asked to supply 3 reliefs of 50 men each to work in 3 hour shifts from
7.30 p.m. to 4.30 a.m. We had to build a Communication Trench

from our lines to B. Com. Trench. I was asked to take the first relief, which was by a long way the hardest, for I had to decide just where to put the Trench and plan it out. We had to make a trench of 20 foot half circles. We returned at 10.30. Gerald Harvey had charge of the second relief from 10.30 to 1.30 a.m. Bullets had been flying around quite freely all night, and at 1.30 when Harvey was preparing to come in, a spent bullet got him in the shoulder. It was not at all a serious wound but would necessitate his removal to the Hospital Ship which when they had sufficient number on board would proceed to Alexandria or some other place. As it was, he was going to be laid off for a short time owing to an affection of his teeth or rather of his gums which were decaying, & only the day before he said to me that he was annoyed at having to be removed from his Coy. for that sort of thing and that he would much rather it was for a wound. So he had his wish granted, which has probably pleased him.

Weather: Fine & warm, the nights too being particularly so.

SUNDAY – OCT. 3RD/15
Strange to say we had nothing to do today – which we availed of as a day of rest. In the afternoon I went to Ordnance Depot (3 miles) to get something from my valise and at the same time had a glorious swim in the Bay. The water was glorious.

Weather: Fine & warm.

Received box of 4 apples & chocolate from Aunt Nell.

MONDAY – OCT. 4TH/15
This is the anniversary date of our leaving St. John's a year ago, having left on Sunday, Oct. 4th at 10 p.m. We have received orders to be ready to go into the trenches again tonight, and are therefore spending the day in making preparations such as getting all supplied with "Iron Rations" (a certain quantity of Biscuits, a tin of Bully-Beef, and a small tin containing Tea, sugar and two pieces Meat Soup Extract), water, ammunition, etc.

Left for the firing line at 7.30 p.m. arrived at 9 p.m. Was on watch nearly all night & had no sleep.

Weather: Fine & warm.

TUESDAY – OCT. 5TH/15
Busy day in D. Coy's portion of the firing line today. We have a very large Sap [tunnel or trench] which is not nearly completed. We were enlarging and improving it. Owing to Harvey having gone to Hospital & Knight being sick, I had to spend more than half of the night on watch again.

Weather: Fine & warm, day and night; the nights are really glorious being bright starry and devoid of flies.

WEDNESDAY – OCT. 6TH/15
The men were occupied all day in improving the Trench and on various other fatigues [working parties]. From 3.30 to 4.30 p.m. some of the Turkish Trenches and positions were very heavily shelled by the Navy and Field Guns, principally the former, which were only about 4 miles away. The marksmanship was wonderful and the effect terrible to see. The Gun work and effects was certainly the greatest we have yet seen; when the High Explosive Shells (which were freely used) exploded, they threw up tremendous piles of earth, which went hundreds of feet into the air and covered an area of almost equal distance. For the first time of seeing it was really awe-inspiring, and yet they say that this is nothing to what we will see when the Guns bombard for a whole day sometimes for the purpose of breaking up their trenches, smashing barbed wire entanglements, and generally disorganizing things, immediately preceding an attack by our side.

I spent tonight out in the open with a sapping party of 30 men. We were connecting our portion of the new firing line with that of the Essex Regt. Owing to the bombardment of the afternoon the Turks were evidently afraid we might make an attack, for they kept bullets flying very freely, and as in this particular place the Turkish trenches were only 60 to 70 yds away we had many very narrow shaves, for bullets struck and passed within a foot of us and sometimes it was only a matter of inches.

During the night one man must have come in contact with a buried body, for the smell was terrible and two or three men were made sick. It was too dark to see, for this spot was in the shade of a big tree. We all got safely into our trench at daylight.

Weather: Fine & warm, day and night.

THURSDAY – OCT. 7TH/15
Received 3 letters from England and one from U.S.A. today.

I spent this afternoon in arranging and making preparations with the Essex Regt. for the connecting of their sap and our sap. After dark spent 2 hours in the vicinity of the connection with Lt. Butler and our working party. At midnight I went out in charge of our working party & came in at daylight (4 a.m.) Of course, as usual we had very many narrow escapes.

Weather: Fine & warm.

FRIDAY – OCTOBER 8TH/15
Received a letter from U.S.A. Spent most of the morning & part of the
afternoon with a working party in the sap. Had to return in the afternoon owing to having been fired upon several times by either the
Essex or London Fusiliers. Two shots very nearly hit me – one striking the bank only a couple of inches from my head. At 7 p.m. commenced a thunder lightning storm – the lightning being the brightest I
have ever seen. At 8 p.m. the rain came and fell very heavily for 2 or
3 hours. It then cleared up a bit but came on again and finally
cleared up about 1 am. It was pitch dark, and it was extremely difficult to move around. We had the men all "Standing to" as we were
afraid the Turks might seize an opportunity such as this to make an
attack. About 11 p.m. we "stood clear" and put up the night sentries,
but instead of the usual one in three, we put up two in three. So what
with the rain, wet clothes & blankets and increased duty, we all spent
a very disagreeable night.
    Weather: as above stated.

SATURDAY – OCT. 9TH/15
Spent this morning in making preparations for changing places with
A. Co., who were to relieve us.
    We moved at noon to A's lines which was the old firing line, 100
yards or so in the rear. The afternoon was spent in getting settled in
our new position. At night, as usual, we supplied several fatigues.
    Weather: Fine & warm, as before the rain storm.

SUNDAY – OCT. 10TH/15
Spent today resting, more than anything else. Capt. March & I spent
a couple of hours with Major Drew in the afternoon, making plans
& arrangements for the forming of the sap on our present right to
that on our old right, thus completing our portion of the new line. At
night (8 p.m.) I went out with the Major, to show the Hants Regt.,
who were finishing the sap, what and how we desired the work done.
When we first got there we met Lt. Ellis [NI] of the R.E. [Royal Engi-
neers] who was with a Sergt., laying a rope along the route to be
taken. About ten minutes before, the covering party for the working
party went out to take up its positions – the working party had not
yet gone out. About ten minutes after the Major & I had been talking to Lt. Ellis his Sergt. came back saying his officer had been shot
by one of our covering party – accidentally. We sent for a stretcher
and as the Trench was very narrow and full of sharp turns, the
stretcher had to be carried at arm's length over the head. As the

stretcher bearers could not manage it, Ellis being a heavy man, I took one end and Capt. Alexander took the other. I went all the way to the Red Cross Hospital with the Party – over 3 miles. The bullet went thro' the left side of his stomach about 3 in. from his side, and as he was quite conscious up to the time we left him 11.30 p.m. – 4 hours after the accident – we thought he would eventually be O.K., but heard later he died during the night.

I did not get back to the firing line until 1 a.m., & had to go out to the sap but soon got away from there again & got to bed about 2.15 a.m.

MONDAY – OCT. 11TH/15
Were informed this morning that we were to be ready to move out tonight. I was Orderly Officer today, so my duties took me over the lines held by the whole of our Battalion. I had to see that they were all clean and in order, but beyond this I had practically nothing to do. In the afternoon we were told we were not leaving the trenches until tomorrow night. I did not get to bed until 2 a.m.

Weather: Cloudy, Cool with a little rain about noon.

TUESDAY – OCT. 12TH/15
Spent today in making preparations to leave the trenches. I went out in the morning to the second line of resistance – 3 miles or more away, which we were to take over from the Worcesters, to make arrangements for our Company moving in & to take them over from the W. After dinner I went out again with some men who took out the officers outfits etc. At 7 p.m. we were ready to move out and the Worcesters arrived almost immediately, but were not ready to let us leave until 9 p.m. We then had many delays in the trench on the way out thro' troops moving in &c., and did not reach our place until midnight. Just before reaching our destination one of our men, Ezekiel [NI], got hit in the thigh by a spent bullet. As we did not have a stretcher with us, I took the stretcher bearers back to where he lay after we had reached our trench as they did not know. I then carried the stretcher with the stronger of the stretcher-bearers and we took him to the nearest field hospital which was, fortunately, only a half mile or so away. Capt. March and I were the last to turn in tonight and that was at 2 a.m.

Weather: Fine but slightly cooler during the day with a few drops of rain about 7.30 p.m.

WEDNESDAY – OCT. 13TH/15
In yesterday's notes, I omitted to state that whilst we were "standing to" waiting to be relieved by the Worcesters, a bomb or grenade was thrown at our trench and burst just in front of my platoon, and quite

Filling sandbags for their
position at Suvla Bay, 1915

Newfoundland trenches,
Suvla Bay, 1915

near to where I would have been standing had I been there, but as
luck had it, I was a little further down the trench. I was on my way
back and arrived at my place less than a minute after the explosion.
One of my Lce. [Lance] Cpls., Tucker, was struck by a piece of iron in
the chest. He was taken to hospital, but I do not think it is serious.[19]

We did no work today, but after dusk some of our men were put to
work cutting down some brushwood just in front of our new trench.

Weather: Fine & warm during day but quite cold at night.

THURSDAY – OCT. 14TH/15
Today all men were at work widening, deepening and generally
improving our trenches, making them suitable as our second line of
resistance in case of a reverse.

Weather: Fine as yesterday.

2nd Lieutenant Owen Steele,
Suvla Bay, 1915

FRIDAY – OCT. 15TH/15
Continued our present trench further up over the hill to our left.

I had a lot of officers Mess goods to assort and allot to the several Messes. At 2.30 p.m. we had an O. Mess Meeting, which I had
asked the Colonel's permission to have. I succeeded in getting clear
of my job as Secty. Treas. & acting President – in fact I am the only
member of the Committee left here. We cancelled our arrangements
with the Civil Service, which was costing 4/- per officer per day.
(A week later we find by letter from Timewell that the cost 6/ ½ per
day).

Weather: Fine, and generally as yesterday.

SATURDAY – OCT. 16TH/15
Continued the trench to the left as yesterday. Another of D. Coy's
men – Gowans[20] – was wounded today, by a bullet from a shrapnel
fired at a Turkish (or German) Aeroplane which passed close & low-
down over us today. The bullet entered Gowans neck and he was
taken to hospital – we do not know yet whether it is serious or not.

Quite a lot of shelling was done this evening by our guns, and we had very scant replies from the Turks.
Weather: Cloudy and cool with a little rain at night.

SUNDAY – OCT. 17TH/15
Service was held at 9.30 a.m. & Holy Communion at 10.30 a.m. in C. Coy's lines to which four men from each platoon in D. Coy. were allowed to go.
The remainder spent the morning working at the trench, improving, etc. as before. Were unable to work during the afternoon owing to rain. It was showery all afternoon.
Weather: As above.

MONDAY – OCT. 18TH/15
Today was occupied with further Trench Extension work.
Weather: Fine except occasional showers – & cold.

EXTRACT FROM LETTER DATED 18 OCTOBER 1915.
Jim and I are both well which is something to be thankful for, when you consider that nearly three hundred of our Regiment have now gone to hospital, among whom are Capt.[Walter Frederick] Rendell, and Lieutenants [Herbert] Rendell, [Cyril] Carter[21] and Harvey who have been wounded. But the most common complaint is Diarrhoea with pains in the "tummy," which has been troubling our boys since arriving in Egypt. Another thing that has troubled everyone is that pest of the trenches; but am pleased to say I have not had one yet, although I have "hunted"[rats] several times. I believe I owe my freedom to the fact that I have followed the "fresh air cure." I am still wearing shorts (and they are quite slack) and keep my sleeves rolled above my elbow. Our doctor is the nearest to it for he also wears shorts. This idea allows plenty of fresh air around the body. However, sickness has created the greater havoc for we have now only 760 or 770 out of our original 1070 or 1080. It is reported that a Draft of two hundred men and four officers have been asked for from Ayr, (our reserve). My platoon has only twenty three active men and I had originally over fifty. However, everything is going dandy we are getting a great experience.

TUESDAY – OCT. 19TH/15
Heard today that we were again to go into the Firing Line tonight; accordingly day was spent in making preparations for moving in tonight. In the afternoon Capt. March & I went in to look over the lines our Company was to take over, to see everything was in order,

and to get any details necessary as to work being done by both us &
the Turks, etc.

At 6.30 p.m. we started for the Trenches and D. Co. relieved the
Coy. of the 1st London Royal Fusiliers which were in the Support
Line, arriving there at 7.30 p.m. By 8.30 p.m. everything was settled
away for the night. Just then we received an order from Battalion
Hdquarters to supply two fatigue parties for work in the firing line
immediately. I had to go in charge of one and we worked until 1 o'c.
when we returned and turned in until "Stand to" – 4.30 a.m.

Weather: Fine, though cloudy and cold.

WEDNESDAY – OCT. 20TH/15

Today was wet & cold and consequently very little work could be
done.

Weather: As above and very cold & wet at night.

THURSDAY – OCT. 21ST/15

We supplied one or two small fatigue parties today though we had
very few men as most of our "effective" men were helping to fill the
firing line for "B" Coy, who, like ourselves has a large number of
men on the sick list.

One day, recently, some 180 men attended the morning sick parade
– yesterday there were about 150. Jim[22] has not been feeling well
lately but is still at his work. I have been quite sick for over a week
now, in fact, sick enough to go to Hospital, but am still sticking to
my work, for I do not want to give up until I really have to.

Weather: Still wet and cold.

FRIDAY – OCT. 22ND/15

This morning we received orders to relieve "B" Coy. in the firing line.
At 11 am the Company moved up. This moving was really a farce,
for, as was the case with "B" Coy. whom we were supporting, we
had only about 75 effective men to put in the firing line which
required over 200 men and about half of these men were already in
the firing line having been sent there to replace "B" Coy's sick. As by
this time we had more men sick and likewise B. Coy., we had to take
practically the whole of B. Coy's effective men to fill the firing line. It
really meant that the only difference created by this change was the
exchanging of places by the Officers. I was feeling particularly unwell
today and as I had been on watch from 2 a.m. to 6 a.m. – shortly
after which I "turned in" – I did not get up until 11 a.m. and consequently went into the firing about noon. I immediately went on
watch and did not get relieved until 8 p.m. when I lay down to try to

get a little sleep as I had to go on watch at 10 p.m. I was unable to
sleep, owing to the pains in my "tummy" being so bad, and at 10
p.m. went on watch again without having slept.

Bert Tait returned for Duty today – we had left him at Cairo suffering from Fever of some sort. A few men who had been left at
Mudros, etc., also came.

Weather: Wet early morning but cleared up quite fine before noon
– changing again to showery and cold & wet at night.

SATURDAY – OCT. 23RD/15
Came off watch again at 1 a.m. and managed to get a little sleep
until 4 a.m. when I took my watch again. I was feeling very poorly
today but decided not to give up yet. I did not come off watch again
until 1 p.m. when I was off until "Stand to" (6.30 p.m.). It will thus
be seen that out of 25 hours I have put in 20 hours on watch or on
Duty. I hardly know how I stood it seeing that I felt almost sick
enough to give up and being compelled to obey a "call" at least *once*
in every hour. However I had a good sleep in the afternoon. Had
practically nothing to eat today, in fact, have eaten very little for the
past fortnight – sometimes a slice of dry unbut[tered] toast for breakfast, a little rice for dinner and say, a little bread & milk for tea.

Weather: Fairly fine with a few showers – and cold.

SUNDAY – OCT. 24TH/15
Nothing unusual occurred today. Jim, who has also been rather
unwell lately, was worse last night and this morning attended Sick
Parade. He was given "M & L.D." (Medicine & Light Duty) so went
back from the Firing Line for the Support Line to get a rest. I was
feeling as bad as I have yet felt, but as it was expected that we would
move into the Reserves for our few days rest about Tuesday I still
endeavoured to try to stick it until we left the Firing Line. I had half
a slice of toast for Bkft., nothing for dinner and a little rice for tea.

Weather: Raining nearly all day and cold, particularly at night.

MONDAY – OCT. 25TH/15
I was off watch this morning from 7 a.m. to 1 p.m. so went to "bed"
and remained in until noon, having breakfast – a little brewis[23] – in
bed. My watch this afternoon was from 1 p.m. to 7 p.m. and it was
as much as I could do to stand but I managed it.

I had only four hours to do on watch tonight. – 11 p.m. to 3 p.m.
Capt. O'Brien and Lt. Churchill[24] both offered to do it for me but I
said I would manage it. However when I went to "turn in" at 7.30
p.m. I felt very bad indeed, and to add to my troubles I was now

suffering from a sore throat. I was practically unable to eat anything all day and could not swallow even my spittle the pain was so great, and the outside right [right side of his throat] was very much swollen. Lt. C. again offered to do my watch and as his turn was on at the same hours as mine he said he would attend to my portion of the line at the same time as his; so after a while I agreed. I then wrapped my throat up with something warm and turned in, but was unable [*sic*] to get very little sleep owing to sore throat & stomach pains. However by breakfast time I felt much better for the rest.

Weather: Fairly fine with a few small showers – cold.

TUESDAY – OCT. 26TH/15
Received notification that we were to be relieved today by the Londons [1st Bn. The City of London Regiment], at 11 a.m. so the morning was spent in making preparations.

We moved out at 11 a.m. but this time did not go to where we were last time, but were sent to a much nearer place. It was a new trench about 400 or 500 yds behind the Firing Line. The afternoon was occupied in getting settled in our new quarters. The Officers quarters were not much of it, so I decided to have a new one made for myself tomorrow.

Weather: Cloudy but no rain and a little warmer.

WEDNESDAY – OCT. 27TH/15
We started more Trench Extension and improvement work and the Officers took turns in superintending the work. I had two men making my "Dug-out" for me and felt so much improved that I did quite a lot of work myself at it.

Weather: Fine with alternate cloudy periods – warmer.

THURSDAY – OCT. 28TH/15
Continued work as yesterday and had my "Dug-out" finished. I had it made very comfortable and cosy as we usually spend a week in the place we go when we come out of the Firing Line. I now consider my "Dug-out" one of the best and most secure from rain in this district. I have room for my bed with room further in for my Kit Bag and Full Equipment and have also a Candle shelf and two other shelves to accommodate books, toilet articles, etc. There is plenty of room to dress inside. I have even a door, made of a brim sack. The only regret is that we will not stay long enough in this place to reap the full benefits.

Weather: Fine & warmer occasionally cloudy.

FRIDAY – OCT. 29TH/15
Nothing unusual today. Continued work as stated above. In my spare hours I wrote one or two letters and a few P.C.'s.
Weather: Very fine and very warm – cloudless skies both day & night – in fact ideal summer weather. Flies are now more troublesome than they have been for a quite a while past.

SATURDAY – OCT. 30TH/15
Work as stated above still continued today. Some heavy shelling was done today by both sides continuing throughout the day. On Wednesday past I omitted to state that just before midday the Turks fired what must have been a couple of hundred shrapnel shells – nearly all in one locality on hill 971 – in about a quarter of an hour. It was the most furious shelling that we have witnessed since coming here.
Shelling is going on almost continuously here and one cannot help wondering why one side or the other does not make an attack. But I suppose there are good reasons on both sides.
Weather: Another perfect day in all respects, except for the flies.

SUNDAY – OCT. 31ST/15
Worked at the Trench again today. Nothing to relate.
Weather: Perfect in all respects.

MONDAY – NOVR. 1ST/15
Did not work at all today owing to the fact that we are to go into the Firing Line to-morrow morning.
Wrote a few letters, and in the afternoon went up to get particulars etc. of, and make arrangements for the taking over of our portion of the line.
The Officers of the First London Royal Fusiliers, whose Battalion we were to relieve, asked me to stay to Lunch – as I left our lines shortly after noon, – so I stayed, for we had previously entertained one of them.
Weather: As yesterday.

EXTRACT FROM LETTER DATED 1 NOVEMBER 1915
Regarding our wanting anything; There is nothing we need, and in any case, if we asked for anything, say now, you would not get the letter until about December the 5th or 6th; and if you despatched it say December the 12th or 15th, it would be about the end of January before we would get it, (for parcels take much longer to get to us than letters) and we do not know where we shall be by that time;

then, again, the expense of sending anything all the way here would be more than its worth. I brought all my warm goods with me although practically no one else did, as we were told it was warm where we were going and we would not need them, which I ignored, as I thought that if we did not need them and if they were in the way, I could throw them away. I have even my long rubbers with me. If at anytime you send us anything, especially anything in the eating line, such as chocolates, Milk chocolate, Cake or anything like this, be sure you put it in something very secure, such as a tin box. It would even be advisable to have it soldered, for many a chap has received a tin box with practically nothing left in it.

This Trench Warfare is really very monotonous, not comparable to "Guerilla" Warfare, as that of the Boer War, where one would get lots of excitement. However, we may get excitement yet.

The past week has been gloriously fine, and I hope the report that November is the finest month of the year here is correct. They say, November is noted for its wonderful sun-sets here, and we have seen some the finest I have ever seen.

TUESDAY – NOVR. 2ND/15
Left for the Firing Line at 10.30 a.m. relieving the Londons in due course.

Everything was very quiet today and nothing unusual occurred.
Weather: Fine & warm.

WEDNESDAY – NOVR. 3RD/15
Nothing out of the ordinary occurred today – just the usual occasional bullet and occasional shell passing over each way.

About 9 o'c. p.m. I had to go out in front of our trenches with 12 men & 1 Sergt. to cover the Royal Engineers who were putting up some additional barbed wire in front of our line. We had to go about 100 yds or more in front of our trench which meant 80 yds in front of our barbed wire. We heard nothing very serious beyond a sniper or two who appeared to be only about 20 yds or less from us.

When we came in Capt. Wilson C.R.E. [Commanding Royal Engineers], who was in charge of the R.E., told me that I managed my Covering Party very well and that it was quite satisfactory.

Weather: Fine & warm, excepting for an hours rain in the morning.

THURSDAY – NOVR. 4TH/15
It was decided to-day to take possession of a snipers post midway between ours & the Turks firing lines. Capt. B[utler]. of "C." Coy., was acting O.C. Firing Line, so it naturally followed that "C" Coy

was given this to do. Lt. D[onnelly].[25] with 6 men and an N.C.O. went out about 4 p.m. and occupied, or rather took up their position in this post which could be easily occupied in the daylight unseen by the enemy. About 7.30 p.m. Lt. R[os]s.[26] went out with out with another 6 men and an N.C.O. as the N.C.O. of the first party had come in wounded and was asked by Lt. D. to ask for re-inforcements to be sent out. It seems that just at dusk three Turkish snipers came stalking up to this post as unconcerned as could be and when quite close were challenged by Lt. D's. N.C.O. The only answer they got was a babble of Turkish; the order to fire was then given to our men and they immediately got the first two men, but the third retaliated and succeeded in getting Lt. D's. N.C.O. in the side of the neck – not seriously though. They claim they eventually got him too. When Lt. Rs. went out with his men when going along a small gully he came into contact with a small party who challenged them saying: "Who goes there" to which Lt. Rs. replied: "Newfoundlanders." The Turkish party – for such it was then said: "Newfoundlands! Allah Allah il Allah," or something similar which they use as a battle-cry, and then commenced firing. Our party of course returned the firing, and the firing continued for quite a while. Lt. Rs's party then returned, for Lt. Rs. and three of his men were wounded. Lt. Rs. was wounded in the arm – but the most serious was Joe Murphy[27] of Mundy's Pond – St. J. He got struck by a couple of bullets and a hand-bomb. Lt. Rs. then said a large reinforcement would be needed to hold the post in case the Turks endeavoured to obtain it. No reinforcement was sent that night, for no one knew exactly where to locate the post, and during that night no one knew whether the first, Lt. D's Party, was OK or not. After midnight every was quite normal again.

Weather: Fine but occasionally cloudy & slightly cooler.

FRIDAY – NOVR. 5TH/15
Early this morning, Lt. D. returned with the information that he and his party were O.K. excepting for two men slightly wounded. It was then decided to place about 30 men with N.C.O.'s in the post and this was accordingly done later in the day, Capt. Rowsell going in charge accompanied by Lt. [Herbert] Rendell. It was generally agreed that, taking into consideration the circumstances, the second party of 6 men which went out should have continued the intention, with which they originally started out, and joined the first party, and that then they would have made a name for themselves.

The day was occupied in getting saps, etc., dug, and sandbags carried out and defensive parapets made by reliefs of working parties. Nothing very serious occurred tonight though there was some bombing and quite a lot of rifle shooting for an hour or two, and then the Turks seemed to give up fighting for it.

Weather: As yesterday.

EXTRACT FROM LETTER DATED 5 NOVEMBER 1915.

Last night whilst 'standing to' at dusk, a Turk came in and gave himself up to our portion of the Firing Line, but unfortunately for D Company, he chose C Company' s end. He lay down by the barbed wire and was then beckoned to by our men. He left his rifle there and came slowly in. He was immediately blind folded and led to the rear , given a feed, and taken to Brigade Headquarters.

A couple of days ago the Colonel was hit in the wrist or hand by a stray bullet. It occurred after dark about a mile from the Firing Line. He went to hospital and we have heard he has gone to Alexandria; they say some bones are broken and an artery severed.

Jack Fox has been sent to hospital and has, I think, also gone to Alexandria – he is sick.

Capt. Carty, O.C. 'A' Company, has also been taken sick and has gone to Hospital.

Lt. [Herbert] Rendell has returned from hospital, having arrived back the day before yesterday.

Major Drew is now commanding the Battalion; Capt. Alexander is acting Major, Capt. Raley is acting Adjutant and Lieut. Wighton is acting O.C. 'A' Company (in Capt. Carty's absence.)

SATURDAY – NOVR. 6TH/15

Everything was as usual today with a little Bombing at night in the vicinity of the new post.

The Commander of the Army Corps of which we form a part passed through our lines this evening and with him was the Divisional General (de Lisle)[28] Brigadier General (Caley)[29] and various Staff Officers. The Army Corps Commander said to us that that was a fine piece of work we had accomplished in connection with the occupying of that new post and said to the Major (our C.O.) that he hoped he was putting in a good report of the affair, and further, that he hoped also that he would mention a few names.

Weather: Fine but very cool at night and flies still numerous all day.

SUNDAY – NOVR. 7TH/15

Work at the Post, or rather hill of Posts, continued without being interfered with except for a little sniping. There are now established on the hill, five posts; one of the posts is now designated "Donnelly's

Post" on account of Lt. D[onnelly]. having first occupied it and the hill is called "Caribou Hill." The Worcesters, the Battalion to our left and immediately behind the hill, have now established five other posts. The object is evidently to extend our trench or firing line along through these posts which will then straighten out our Line.

Weather: As Saturday.

MONDAY – NOVR. 8TH/15

The Turks were a little busier than usual today with their sniping but no damage was done, for in day time all Sentries use Periscopes. At night quite a little bombing and rifle shooting took place at "Caribou Hill;" "D" Coy. had charge of the Hill from about 4 p.m., Capt. O'Brien taking out about 30 men.[30] I would have very much liked to have the chance of getting out there but we have the Firing Line also to look after. Our party landed some very well thrown Bombs into the Turkish Redoubt and Trenches near by. One of the flares sent up for the purpose of seeing if there were any bodies of Turks near and also to throw a bright light into the eyes of the Turks in the Trenches so that they would not see the burning fuze of the hand grenade or bombs coming towards them, set fire to the woods and low bushes midway between the Hill and the Turkish Lines. This fire assumed quite large proportions and lit up the vicinity for at least a mile around. The smoke, unfortunately, was blown towards our position and very nearly drove our men out of our left post.

It was intended that I would go out and examine the barbed wire in front of our trenches to see if it was O.K. and had not been tampered with, at 7.30 p.m., with Sergeant Beth.[une][31] and a private for watching and listening purposes, but as this fire lasted for 3 or 4 hours, we had to put it off until later in the night. We were still further delayed by some bombing being done by "C" Coy. on our right which was drawing quite a lot of rifle fire. It was eventually decided that we would go at 3 o'c. tomorrow morning.

Weather: Fine, and cool at night, but not so cold as last night.

TUESDAY – NOVR. 9TH/15

At 3 a.m. we went out and examined the Barbed Wire and found everything O.K. and no shots were fired at us – we were all armed with rifles and 50 rds of ammunition each.

This morning we received word that we were to be relieved by the Londons at 7 p.m. tonight, and that we were to occupy their present position in the Second Line of Resistance, which was the place we were in the second last time we were out of the Firing Line. At 10.30 a.m. I was sent out by Capt. March to look over the Trenches we were to take over tonight and returned about 1 p.m. The weather was gloriously fine until 3 p.m. when it became a little cloudy and cooler.

At 5 p.m. when Orders came out we were all delighted to read the very complimentary remarks from the G.O.C. concerning the Battalion's work in connection with the occupation of Caribou Hill.

We were to have been relieved by "The Londons" at 7 p.m., but they were late in arriving, and it was 8.30 p.m. when we were able to leave the Firing Line. We walked very slowly out and towards the end of "C-D"[companies'] Communication Trench met a Company of "The Londons" going in to relieve our "B" Coy. This meeting and passing of another Company in the Communication Trench caused a great delay and lots of confusion for there was not sufficient room for the Troops to pass with their packs. Some of the other Company sat down on seats or Firing Steps that were there, others stooped down so that our men could get pass [past] with their packs, others sat down and our men stepped over them; however, our men eventually succeeded in getting past, but all the while, at the particularly hard places, there were plenty of curses, swearing, yells of pain from the men who were being walked on, sarcastic remarks, etc., etc. Others of "The Londons", who were comfortably seated, spent their time in laughing at and making all kinds of remarks to those of their comrades who were not in so fortunate positions or circumstances. However, we eventually reached our destination, the Second Line of Resistance, about 11 p.m. After we had all the men fixed off in their places, we prepared to turn in ourselves, but before doing so had a little supper of Bread and Golden Syrup, for we were hungry after our rather strenuous evening and tiresome walk.

Weather: Quite a glorious day, and night fine but cool. Towards morning we had a shower or two of rain.

WEDNESDAY – NOVR. 10TH/15
We are now in the same place we were in a month ago, and our particular work this time whilst out of the Firing Line for a "rest" is the digging of a Communication Trench from the Second Line, which is our habitation for our week out of the Firing Line, to some distance to our rear, so that if such a thing occurred that we were driven back and had to use our Second Line as a Firing Line, troops could come into it from the rear under shelter. So our men spent the day at this Trench.

Weather: Fine & cool.

THURSDAY – NOVR. 11TH/15
Quite a lot of shelling was done at the right of our Company this morning by the Turks. None of our men were injured; which was lucky, for it was all shrapnel that came over, but a man from "A" Coy. was on the way to our Company with a message and had just reached our lines when he was struck by a few shrapnel bullets, getting two wounds in the arm and one in the knee.

Our men continued the Trench work today.

Weather: Fine at some times and occasionally turning suddenly showery, and cool.

FRIDAY – NOVR. 12TH/15
I have now got a new "Dug-out" made for myself, and have had it made to my own liking and a style of my own – it was completed yesterday. With my rubber sheets rigged up overhead, underneath which is a blanket, I have the satisfaction of knowing that it is entirely waterproof – I have little drains along the sides to take away the water as it runs off the rubber sheets – and further, it is quite warm.

There are many rumours going around continually, as to our moving, and they have been going around for about a fortnight now, but are now getting more persistent than ever. They are such as these: That our Division – the 29th – (our Brigade is the 88th) – is leaving here for Imbros for a rest – an island 4 or 5 miles from here. Others say, to Egypt (Cairo, Alexandria, Malta, &c), to England, to Salonica [Salonika], etc., etc. However, I guess the best thing to do is to "wait & see", but I must say everything, or at least very many things, seem to point in that direction.

Our men continued the Trench work today, and some worked for 4 hours at night.

(Received box of 2 cakes, chocolate & candy from Aunt Nelly today.)

Weather: Fine but several heavy showers at night, but I kept dry in my "Dug-out."

SATURDAY – NOVR. 13TH/15
Rumours are growing stronger each day, and they now say we are leaving here on Thursday next – 18th inst. Nothing unusual today. We continued the Trench work already mentioned, and at night half of our men worked at a portion of the Trench that could not be done in daylight.

Weather: Showery all morning, but fine in the afternoon.

SUNDAY – NOVR. 14TH/15
For a wonder we did not work today, but to make up for it the men
worked for a few hours at night (4 hours). I was on duty, superintending, etc., for 2 ½ hours. Today was very quiet indeed, but quite
interesting nevertheless, for I received about half a doz. letters, which
kept me busy reading & then writing.

Weather: Fine generally with one or two small showers – cool.

MONDAY – NOVR. 15TH/15
Trench work continued today. Once or twice during the morning had
to stop work on account of heavy rain showers. Did not work at all
during afternoon or night on account of going into Firing Line again
tomorrow. During the morning quite a little general shelling was
done by both sides, but from about 3.30 p.m. for an hour or more,
some heavy shelling took place over at Aschar [Achi] Baba, the Warships – some away out just on the horizon – taking the principal part
apparently. Then for an hour from about 4.30 p.m. some of our land
batteries & a few warships heavily shelled a portion of the Turks line
just to the rear of Chocolate Hill and at the foot of 971 Hill. This
shelling was very heavy and continuous and was accompanied by
quite a lot of rifle fire. It was certainly exciting whilst it was on and
made us all feel that we would like to be into it, for things have been
very quiet indeed lately. We do not, nor are we likely to, know the
result or object of the bombardment.[32]

It has been said lately that Lord Kitchener has been over at Imbros
– an island, Main Hdqtrs, I believe, 4 or 5 miles off – for the past
two or three days and some say the bombardment was for his benefit. At night everything was as quiet as usual.

Weather: Showery in the morning, fine in the afternoon. At Tea
time, some heavy clouds passed over the Western part of the sky and
there was some very vivid lightning with occasional slight thunder.
We thought we were just escaping it but a couple of hours later, the
lightning came again and clouds spread over the sky. We then had a
terrific lightning thunder & rain storm – the severest I have seen for a
long time – the result being that everyone was washed out. I alone
was dry, and all my belongings, for I had a "Dug-out" which I made
myself & rigged up my rubber sheets in such a way that I was not
troubled with any rain at all.

TUESDAY – NOVR. 16TH/15
As we were to go to the Firing Line again tonight our men did no
work, but rested and made necessary preparations for moving. At
11 a.m. I went into the Firing, taking the Sergt. Major with me, to

look over the lines we were to take over, after which I was to meet
the Major [T.M. Drew][33] – now our C.O. – Our Coy. was to go
into the Support Line, taking over a portion which really constituted a Forward Support. I arrived back to our present quarters at
1.30 p.m. After having dinner I went, with my orderly, to the
Beach, for the purpose of putting some things that I did not need
in my valise, and to get some things I did need – among the latter
being my long rubbers. About 4 p.m. whilst I was still at the
Beach, another bit of "Straffe-ing" was done by our Artillery in the
same place as last night. It was not nearly as severe as last night,
tho' quite interesting whilst it lasted. I got back at Teatime. I was
very pleased indeed at finding my valise, for over half a dozen of
the officers had had their valises tampered with & goods stolen, in
fact some had even the valise, intact, stolen. My orderly had, some
time ago, broken my key off in the lock, so I had my valise tied up
with cord, so fully expected that if mine was there at all to find
that it had been tampered with, but was pleased to find it
untouched.

We left for our new position at 6 o'c. p.m. and arrived about 7.15.
After getting the men fixed off, we commenced our watches for the
night – mine was the first one, from 9 p.m. to midnight. I then turned
in until "stand to" (5 a.m.)

Weather: Fine and cool.

WEDNESDAY – NOVR. 17TH/15

Practically all our available men were on working details today – in
fact, will be until we either go into the Front Line or back to the
Reserves. At 9 a.m., Major Drew wanted to see me, and asked me to
superintend the construction of a Bath-House for the men, and,
about half a mile away, the construction of firing positions in C.22
Communication Trench, leading to the Firing Line. This kept me fully
occupied all day.

Nothing important occurred today.

Weather: Fine during the day, but a small rainstorm about 8 p.m. –
Cool during the day but cold after rainstorm.

THURSDAY – NOVR. 18TH/15

Was Orderly Officer today, but as the Duties appertaining thereto,
are not very multitudinous here, I still looked after those two jobs
given me by the Major yesterday. At 11 a.m. went, with the (Dr.)
[Lamont Paterson], around all the lines occupied by our Battalion, to
see if they were clean and that everything was in order. We had a
"D" Coy. man killed this morning. His name was [Joseph] Hynes,[34]

(of the Southside[35]) and he was engaged in some work at the Firing Line.
Weather: Fine & cold.

FRIDAY – NOVR. 19TH/15
Continued Bath House & Firing Steps as above.
Weather: Fine & warm.

SATURDAY – NOVR. 20TH/15
Still continued above mentioned work today but was disappointed in my hopes to finish the Bath House today. Rec'd parcel from Mrs. W.M.[William Martin]
Weather: Cloudy & cold.

SUNDAY – NOVR. 21ST/15
Finished Bath House at noon & continued Firing Steps.
Received parcel from home, as did also Jim – both in excellent condition.
Weather: Fine & warm.

MONDAY – NOVR. 22ND/15
Bath House was used today for first time, about 35 using it.
Continued Firing Steps & finished.
Weather: Damp & Cold.

TUESDAY – NOVR. 23RD/15
Today the Major gave me a new job – to make a wash house, a place for the men to wash their faces and hands. Over a hundred men bathed.
Weather: Fine & warm.

WEDNESDAY – NOVR. 24TH/15
Continued above job
Weather: Fairly fine with one shower of rain – a little warmer.

EXTRACT FROM LETTER DATED 24 NOVEMBER 1915
By this time, you will no doubt have heard of the exploits of C Company. Lieut. Donnelly was awarded the Military Cross, Sergeant [Walter M.] Green[e] the D.C.M. and Private [Richard E.] Hynes[36] the D.C.M. I cannot give any details; however, I expect you will hear things officially.
We are to be in the Firing Line this time for a fortnight, and will

then be moving out to the Second line of Resistance for a REST – but what we get then is better described by the word WORK.

Victor asks for particulars of our present officers. I cannot be more explanatory then to mention their names, which are as follows:

A Coy. – Capt. Carty, Lts. Brighton [NI], Shortall[37] & Stick.

B " – Capt. Alexander, Lts. Tait, Nunns & Clift[38]

C " – Capts. Bernard[39] & Rowsell and Lt. [Herbert] Rendell

D " – Capts. March & O'Brien, Lts. Steele, Knight & Reid.[40]

and the following:

Major Drew, Capt. Raley, Lts. Summers, Frew[41] and Butler. Lieut. Windeler[42] is now Brigade Machine Gun Officer, so is really out of our Regiment. The officer who held that post before is sick and has been sent to England.

THURSDAY – NOVR. 25TH/15

Finished above job today. The Major was very pleased with my jobs & paid me various compliments, and said he would show it to the Staff when they came around – and he did, and they asked who had done the jobs and they were all exceedingly pleased. Lt. Wighton killed instantaneously visiting listening patrol. Bullet thro' brain.

Weather: Fine & warm with a shower at night.

Was kept generally busy today. Tonight I had to spend 4 hrs in the Firing Line. I left the firing line at daylight. Lt. Shortall was wounded just afterwards.

Weather: Fine & Cold.

FRIDAY – NOVR. 26TH/15

I spent today in the Firing Line to help out "A" Company as they had only two officers, Capt. Carty & Lt. Stick.

I was out of the Firing Line tonight and Lt. Knight was taking a 4 hrs watch there. About 6 o'c. it commenced to rain. For 4 hrs it thundered lightninged & rained the heaviest, by far, that I or anyone else in our Regiment had ever experienced. Tons of water came rushing thro' the Trenches. Men lost blankets, rifles, equipment, etc., etc. Sides of Trenches were washed away, fell in, & became actually mud as soon as they got wet. Parapets fell in, were washed down, trenches were filled with debris of every kind. The rivers of water running thro' the trenches and through the drains we had made were filled with floating blankets, equipment, etc., etc. In fact waters were far beyond description. At 10 o'c. it stopped raining and cleared off; then one could get a more vivid idea of the damage done and being done. I cannot attempt to describe its various details. The floor of my dug-out was a foot and a

half above the bottom of the drain, yet the water was a foot & a half
or more deep in my dugout. It all came in in a rush and at a time when
I was keeping it baled out with a cigarette tin nicely, some water having found its way through the roof and down the sides. Up to this
time, I had kept all my things dry, but most of it now got wet.

All our Headquarters were completely washed out, all Officers
dug-outs, etc. All Orderly Room papers & books were washed away
and everything in fact was demolished.

There was no sleep for anyone tonight, and when morning came
confusion and disorder reigned everywhere.

Weather: Quite evident from above.

SATURDAY – NOVR. 27TH/15
I was Orderly O.[fficer] today. Everyone getting things in order
    Weather: Snow and hail storm tonight.

SUNDAY – NOVR. 28TH/15
I was ordered to the Firing Line this morning & remained there all day
& night. Of course everyone had wet feet & wet clothes all the time
and no way to dry them. However everyone stuck it very bravely considering the extreme conditions. The reason I had to take the Firing
Line this morning was because Capt. Carty (O.C. "A" Company) was
on a Court Martial Case and had to go out to the Beach – 3 miles –. I
was therefore "O.C." "A" Company for today.

Capt. Carty must have had a very hard time getting out, for the
trenches were half filled with water everywhere and it was very cold,
in fact, in most places the water was frozen to a thickness of from
1/4 to ½ inch, then again, when Carty went he was not feeling at all
well. However, no one was surprised when, up to 5 p.m., he had not
returned, as it was expected he would not be able to return and that
he would go to Hospital, which was near to where he had gone. At
"Stand to" (5 p.m.) the Adjutant informed me that, as Carty had not
returned, I was to continue o/c "A" Company until further orders. I
was in the Firing Line again tonight and had the pleasure of being
O.C. of that portion of the line held by C. & D. Coys. Nothing
unusual occurred tonight, but it was intensely cold, and what with
wet feet & legs half way to the knees it was a very uncomfortable
night. It was impossible to keep or even get oneself warm, for one
was in mud & water all the time and space for moving around was
very limited. Everyone – mostly myself – was very surprised at my
being given such an important responsibility as O.C. Company Firing
Line, a post, until tonight, always held by a Captain.

    Weather: Dull and very cold.

MONDAY – NOVR. 29TH/15
Was out of the Firing Line today. After breakfast the Major himself
told me he wanted me to take over the command of "A" Company
until further orders. He asked me to try to reorganize the Company
as they would be coming out of the Firing Line this evening, – we did
not know whether we had any men missing; rather we knew a couple
had gone out to the Beach but were reported as having gone to Hospital. I eventually found that 9 had gone, but 3 returned later and the
remainder, we subsequently found, went to Hospital. What with the
Company work entailed by reorganization I had an exceedingly busy
day.

Carty returned during the afternoon, having remained at our Regimental Dump for the night. When he asked Major Drew if he would
take over the Company, the Major said, no, but to leave it to me as
he had asked me to reorganize the Coy. & that I was doing well with
it and that he did not want me interfered with in any way until I had
finished. He told Carty to go to his Dugout and get all the rest he
could, as he was not feeling at all well. Carty himself told me this &
said he was only too pleased I was able to do it.

Weather: Cold but fine.

TUESDAY – NOVR. 30TH/15
Spent today in drawing up new Platoon & Company Rolls and
recording in my Field Note Book a complete list of the men, their
physical condition, and all deficiencies of Rifles, Bayonets, the various parts of Web Equipment and outside & inside clothing etc. This
took up all day. A summary of this list I then handed to the Quartermaster.

All the well men of my Coy. had to go to the Firing Line tonight
and also of "D" Company and I myself had to go in and act as O.C.
of the Firing Line thus occupied. "D" was asked to detail also a Subaltern so I had Lt. Reid with me. We had only 36 men and 10
N.C.O.s to hold the line for this was all the well men we could get,
whereas we used under normal conditions to have a total of about
100. Everything went O.K. until 5 o'c. in the morning when one of
the sentries was shot in the head death being instantaneous. I was at
his side in 3 mins, but he was dead. Jim, who was Sergeant in that
vicinity, was nearby and helped place him on the firing step immediately it happened & then sent for me – he says he was dead even
then. The Turks did quite a lot of sniping tonight, in fact since "The
Flood" they have been doing much more than usual during the
nights.

Weather: Fine but cold.

WEDNESDAY – DECR. 1/15

Was surprised – in one way – to hear this morning before being relieved of the Firing Line, that the Major was going to Hospital. He left about 10 a.m. and Capt. Carty immediately took charge, though he himself was not much better. Carty then told me he wanted me to continue in Command of "A" Coy. until they got things straightened off.

Continued fixing up various matters in connection with "A" Coy. today. At 10 p.m. our first draft 93 men and 4 Officers, – arrived and landed in the Beach and at midnight they reached our Trenches. Capt. Raley sent Paterson [Patterson], Lt. to my Dugout to see if I could make room for him. Just before that, [George] Taylor had come in. However I made room for the both of them – this was about 3.30 a.m.

Weather: Very frosty but fine.

THURSDAY – DECR. 2ND/15

Was fully occupied with "A" Coy. work all day. Had to put a change of men in the Firing Line tonight. We had a full line tonight having almost a hundred men & N.C.O.s. I was O/C of the Line and had with me Lts. Knight, Bartlett[43] & Strong.[44] Everything O.K.

Weather: Cold & damp during day and raining slightly all night – very cold.

FRIDAY – DECR. 3RD/15

About 10 or 11 a.m. Capt. O'Brien, Second in Command of "D" Company took over command of "A" Coy. and I returned to my own Coy. Capt. O'Brien had with him Lt. Stick and one of the new draft Lt. Strong. The other 3 officers of the draft were distributed as follows: Lt. Taylor to B. Co. – Lt. Patterson to "C" Coy. & Lt. Bartlett to "D" Coy. About noon Capt. Carty, now acting O.C. Battalion, went to Hospital. Even when he took over the Battalion he was not fit to do so, for he had been unwell for several days. This duty of O.C. Battalion now devolved upon Capt. Alexander.

Immediately Carty went to Hospital Alexander took charge. ·

Weather: Fine all day, but at 6 p.m. the following message was received from Brigade Hdqtrs: – "H.M.S. *Cornwall* wires every appearance of heavy rain storm tonight." All then made preparations for another "Flood" for the sky was indeed very cloudy, but all were pleased that no rain whatever fell.

SATURDAY – DECR. 4TH/15

Well today is just a week after "The Flood" and things are now almost normal again. Still there is quite a lot of work in the way of cleaning up the trenches and making repairs. As will perhaps be noticed, I have written all that has occurred since just before "The Flood" almost all at one time, not having had time since that never-to-be-forgotten experience of Friday, Novr. 26th. I will now endeavour to mention briefly a few details of our and others' experiences.

Our Regiment is stated to have come out of the affair the most satisfactorily by far. Of course some Regiments or Battalions were not very greatly inconvenienced by the storm such as the Essex, who were on our left, for they were higher up on the hill. We got it very bad indeed and so did the rest of the 88th Brigade, who were on our right. The Worcesters, The Hants, and The Londons were on our right. The latter three lost some 100 men by death from exposure during the couple of days frost which followed the flood. They had several hundred sent to Hospital with frost burnt feet, sickness, etc. One of Worcester Officers told me four days after the storm that he had the previous day, been down to the "Block-House" (about a mile to our right) in the lines held by the 86th Brigade, where he had seen many men lying dead in the trenches and being walked over. He also saw fully 30 men sitting up on the firing steps exactly as they were at the time of the storm, frozen to death. There was so much work to do in clearing away and burying the dead that these bodies had not yet been attended to. He said it was a very gruesome sight indeed as was also another sight he had seen. We had recently built some winter Dugouts, made to hold about 8 men lying down. Well an officer of one of the 86th Brigade Battalions had evidently gone into one of these and taken some men with him to the number of about 30. These are without roofs, being unfinished, so the result was that the officer and every man froze to death. Another officer told me he was standing in the doorway of his dug-out, when he saw in the river of water flowing down just outside the Trench two dead mules, one live mule, two dead Turks and many boxes and large pieces of debris pass down altogether.

On the night of the Flood our firing line & support Trenches were about 3 feet deep nearly everywhere and many places men had to walk in water up to their waists; everyone, of course, was wet to the skin and unable to dry themselves, then when the frost came it tried us all to the limit and all suffered severely, but thanks to the general hardiness of the Nflders, not one death resulted in our Battalion, and I know there were very few Battalions indeed around here which could say that: Of course sleeping, for the majority was out of the question, for many of the men had lost blankets & rubber sheets and quite a few had even lost their great coats. So fires were made without any thought to the fact that we were showing our exact position to the Turks, but the Turks were, if anything, worse off than we and they also had fires burning everywhere. All these big fires were just in rear of the Support Line, and men on both sides wandered around and stood around the fires with a wanton and utter disregard of the other side. Neither side bothered about the other in the least, until the severity of the frost began to lessen. Our firing line then got many wandering Turks, and they got a few of ours. Owing to the Communication Trench being 2 & 3 feet deep with water in most places, our Ration Parties had got into the habit of walking in the open with the result that the Turks noticed it and eventually got a machine gun on to the track followed and also sent in a few shells. We lost rather heavily in this for two days work, for we lost about half a dozen killed, and about a dozen wounded. Most of the Officers too lost most of their kit. I will never forget the look of most of our men after the first & second night's frost. It reminded one of the "Greenland Disaster"; the mens' faces were nearly all as black as niggers from sitting over the smoky fires all night and what with their white eyes and woe-begone looks they presented a really terrible sight. One was fully expecting to find [them] lying dead from exposure, but, wonderful to relate, there was not a single fatality, which speaks well for the physique of our men, for those were really terrible nights. Even several of the Officers who were in the supports – such as Hdqtrs. Staff – were blackened somewhat from sitting over a brazier in one of their Dugouts. I lay down myself for a few hours those cold nights, but was practically unable to sleep on account of the cold. Jim had a cosy place in some woods with his blanket, great coat & rubber sheet – also oil coat. He was allright [sic] in the mornings except for a bit of rheumatism in his hips, which made it difficult for him to move around. I was o.k. because I kept on the rush nearly all the time.

After the Flood we were very short of Rations for a while and had to do without some meals, for want of both food and water. It was with the latter that we had the greatest difficulty, for all the wells were spoilt, the water they contained being exactly the same as the Trench water. The Doctor condemned the water for drinking purposes and the first day we had nothing to drink, and the second day one lot of water for making Tea, but the drinking of the unboiled water was absolutely forbidden, for there were so many dead bodies and rubbish of all kinds around that an epidemic of some kind would have resulted. Even the water that we then had was really only muddy trench water in which we would not have attempted even to wash our hands under normal conditions.

Well, I cannot spare time to write any more but could go on writing for hours there being any amount of material. Suffice it to say that we came out of the terrible ordeal, which has seemed like a terrible nightmare, in a very satisfactory manner and are now as busy as can be repairing the immense damages.

We have sent about 150 men to Hospital most of them being for frost burnt feet.

We have heard that the 86th Brigade lost 200 men by drowning and exposure and nearly 2000 were sent to Hospital. They were on the right of our Brigade. We spent today in repairing drains, trenches etc.

Weather: fine and quite warm.

SUNDAY – DECR. 5TH/15
Nothing particular occurred today – working as usual. (Rec'd 4 letters today.)

Weather: Fine & warm.

CHAPTER FIVE

Evacuation
(5 December 1915 – 11 January 1916)

T he decision to evacuate the Gallipoli Peninsula was first considered
in late September after the failure of the August offensives. Tenta-
tive hopes of a renewed offensive were dashed with Bulgaria's entry
into the war on the side of the Central Powers and the resultant threat
that posed to an already hard-pressed Serbia. That threat compelled
the Allies to shift forces to Salonika where they and later reinforcements would sit idle for the rest of the war. The decision to open up yet
another theatre sounded the death knell of the Gallipoli operation.[1] On
7 December the troops at Anzac and Suvla were informed that they
would be evacuated.[2]

As soon as Anzac and Suvla were safely evacuated, attention turned
to Helles, at the base of the peninsula, which the French still held
despite heavy Turkish fire. The 29th Division, its ranks depleted by
heavy casualties, was assigned the onerous task of covering the French
evacuation. This meant a return to entrenched positions and a rigorous
defence in the face of possible Turkish action against the retreating
Allied forces. The final evacuation was a complex affair that took ten
days to accomplish and required a phased withdrawal from some 20
kilometres of trenches followed by an embarkation from a beach well
within range of enemy shell fire.[3] Then, in two successive nights, the
final garrison of 22,000 men, including most of the Newfoundlanders,

was removed. The British employed every possible device to deceive the Turkish forces of their intention: " men marched round in full view of the Turks on open roads, no tents were struck, the guns fired their regular quantity of shells, the patient mules toiled daily to the lines."[4] It was, as the new regimental commanding officer, Lieutenant-Colonel Arthur L. Hadow[5] later described it, "a very anxious time."[6]

In spite of impressions to the contrary, the scene at the various beaches was far from orderly and there was considerable confusion in the final moments of the departure of the Allied army from Gallipoli. On the night of the 8/9 January the last elements of the covering party, a detachment of the Newfoundland Regiment under Lieutenant Steele, prepared to leave. At 3:45 a.m. the last barge waited at W Beach for Major-General F.S. Maude, commander of the 29th Division, who had returned to Gully Beach to collect his valise. When Maude failed to show up Steele was sent to look for him. The event was recorded by Robert Rhodes James in his book *Gallipoli*: "Lt. Steele, the young Newfoundland officer, was sent to look for Maude. It was pitch-dark, and he knew the main magazines would explode in half an hour. The British trenches were now empty, and Steele felt wretchedly isolated, wandering about in the dark, shouting Maude's name. Maude's party had been held up by having to cut its way through the barbed wire placed across the tracks, but eventually Steele found it. Trundling his bag on a wheeled stretcher, the general arrived at W. Beach just after the commander of the boat had said that he could wait no longer."[7]

MONDAY – DECR. 6TH/15
About Midday (acting) Lt. Col. A.L. Hadow came to take command of the Battalion. Capt. Alexander went back to B. Co. though he is laid up with the malady which is affecting the majority here namely, Trench-feet.

The affection [affliction] is that the feet have a burning sensation, are very sore and much swollen. It is the result of standing in water – which is ice-cold – and thro' being unable to get a change, keeping them [socks and boots] on all day and night. And the feet being ice cold all the time, it being practically impossible to keep, or even get, them warm. Mine have not troubled me, for I have worked hard to keep mine warm and have changed my socks every night having a spare pair with me. The wet ones I dried around my waist, by each succeeding night.

The main work of the Battalion was the improving and lengthening of a drain which is to drain the trenches during the next Flood – (?)

Weather: Fine and moderately warm.

TUESDAY – DECR. 7TH/15

When writing up Thursday Nov. 25th I omitted to mention that about midnight we received orders for all officers to pack and send out to the Beach all their surplus kit. The understanding was that we were leaving the Peninsula within the next few hours. So practically all of us sent away everything except what we were to carry on our backs, but nothing happened and we are still here – it is thought that the Flood or Rain Storm changed all intentions, or at any rate delayed them to such an extent that we have not moved yet.

The same day the Flood occurred that is Friday Novr. 26th consequent to the above order received Thursday night, we sent away the worst of our sick men to the number of 41, which was indeed a very fortunate thing for the Battalion as with these sick men lying on the firing step of the supports our Flood results would probably have been different.

This morning the acting Adjutant, Capt. Raley, went to Hospital, his complaint being Eczema. He was replaced by Tait. I was still in charge of drain grading work all day, which I have been in charge of for about a week now.

At 3.30 p.m. a bombardment of the Turkish lines was made by our land and sea batteries and lasted until 4.30 p.m. We received intimation of it in the forenoon. It was the finest bombardment by far that we have beheld since coming to Gallipoli Peninsula. I had intended getting to the Essex lines just to our left to get a few photos of bursting shells, but at the last moment was unable to get away. I was very sorry too for just about a hundred and two hundred yards in front of their line the heaviest bombardment took place, and the shells there were nearly all "high explosive" (H.E.) and their bursting was indeed magnificent, and would have made beautiful pictures. We do not know of course, just what damage we did, but it must have been tremendous. The object was simply to do as much damage as possible and to terrorize and disorganize the Turkish troops, in which we know we were successful.

Another dozen men went to Hospital again today.

Weather: Fine and warm; just a little cloudy. Night chilly only.

WEDNESDAY – DECR. 8TH/15

I have finished the drain work and today am on some work in our main Communication Trench C-D) [Company] I had a party at work at 6 a.m. to 8 a.m. then 9 a.m. to 12.30 p.m., then 2 p.m. to 4.15 p.m., then 5 p.m. to midnight, when I turned in until "Stand to" (5.45). In the afternoon we received word that it was gathered from prisoners, who had surrendered themselves, that a Turkish Div. to the right of our right front was going to be relieved by another Division, so it was intended to do a little shelling and machine gun firing, so they gave us warning so that we would know what was going on.

About 9 or 10 men went to Hospital again today – they are going slowly but surely.

Weather: Fine and warm but a little chilly at night.

THURSDAY – DECR. 9TH/15
I was very pleased today to receive 7 letters (2 from home), & remember I have not written a single letter or P.C. for a fortnight or more.

Continued work on the Communication Trench today. About 3.30 p.m. my working parties (2) were called in to report to the Quartermaster as were all other working parties, an order having come in to collect all our trench tools and general stores, all surplus ammunition, bombs & catapults, etc., all Officers' surplus rations, company water cans, etc., etc., and send same to Brigade Headquarters. This certainly looks like a move and an immediate one too. Though a somewhat similar order came to us a fortnight ago tonight, but I have heard that the reason nothing materialized then was that things were delayed by the Storm, so I think all will be O.K. this time. There have been hundreds of different rumours during the past month entirely different from day to day, and no one is sorry for this probable move, and the men need a rest too; if they are not soon taken away for a rest there will soon be no Nfld. Battalion here, for the men will eventually all get to Hospital and they are going fast now. Only one man went to Hospl. today, but since the "Flood" and not counting the 41 we sent the morning of the storm we have sent practically 200 to Hospital. This will give some idea of what we underwent during the week subsequent to the storm and Flood.

About 5 p.m. this evening our batteries shelled a small area of Amafarta [Anfarta] Ridge about 200 or 300 yards square and must have sent some 80 to 100 shells – mostly shrapnel – over in less than 3 mins. We are all anxious as to what the next 24 hrs. will bring forth.

Weather: Fine & warm, and really quite pleasant.

A battery near us sent 80 rds over to the Turks in about 1 ½ mins. – a continuous succession of explosions.

FRIDAY – DEC 10TH/15
Silent this morning in the Firing Line. At Dinner time the Colonel (Hadow) sent for me and told me he had another job for me. He said he wanted me to go down and occupy a certain portion of the

Second Line with 30 or 40 men. We were to go down as soon after tea as possible and upon being asked I said I could start at 5.30 p.m. to which he said all right. I then spent about an hour with him making plans and arrangements. Just before leaving him I said: "It looks like evacuation, Sir," to which he replied: " That's just what it is, absolutely." I spent the rest of the short afternoon in making my preparations & at 5.30 p.m. reported to the colonel that I was ready start and at 5.45 p.m. we started. Every man was in "Full Marching Order" & carried a pick and a shovel each. We also brought along two dixies, so with their rifles each man had a full load. We had about 3 or 4 miles to go and I had not been there before so the Colonel showed it me on a map. I had not the least difficulty in finding it though we had to be rather accurate so as to locate the gap in the barbed wire, but what with men with sore feet and one or two men not feeling well, we did not reach our destination until 8.15 – 2 ½ hours. I then fixed the men off for the night and this done went to a certain bridge half a mile away where I had asked the rations [be sent]. On arrival there I found that the man in charge had left the Indian & his mule, cart and our rations for only 2 mins and on going back found the Indian with his charge was not there. These Indians are very restless and as he had been there for an hour when [went] off to look for us himself. As this happened an hour before I got there I sent the Corporal back to our Regt. dump on the beach (where all our rations came from) to get a new lot and bring them to some men who were guarding a well and with whom I then made arrangements, this being half a mile from us in another direction but a place of safety. I then got back to my new lines and after appointing 2 cooks turned in at 11 pm, after a rather strenuous evening
Weather: Fine and warm.

SATURDAY – DEC 11TH/ 15
At day light I went with 4 men for our rations and got them O.K. – On the way back I was in front with a load on my shoulder, as there was more than the men could carry, and one of the men was five yards in my rear when several shells from " Johnny Turk " came over, being fired at a battery near which we were passing. Suddenly I heard one coming that I knew was coming directly for us and as we were in the open had no cover to make for and it was an anxious moment for they were time-fuse shells. Suddenly it landed about four yards from me, but fortunately the fuse was a little late, for it exploded just as it struck the ground and sent a cloud of earth and dust only over me. I certainly had a very narrow escape this time as did also the man in my rear, for if the shell had exploded one second,

even, sooner it would probably have meant an addition to the Newfoundland casualty list.

I had with me besides my orderly, 30 men including a Sergeant & a
Corporal and we spent today in building – or digging – a cookhouse,
latrines and improving a communication trench. In the middle of the
afternoon I commenced a Battalion Headquarters, the Colonel having
been down and asked to do so. He was down about 2.30 p.m. and
went over the whole of our portion of the line – 500 yards and the
other things we were doing and expressed himself as pleased with
how things were going.

About 5 p.m. the Battery at Lala Baba – a few hundred yards away
– gave " Johnny Turks " 80 rds in 1 ½ mins. same as that other battery on Thursday night.

During the afternoon the Turks sent a score of shells out at some
Transports in the Bay without damage, though some went very near.
The Transports immediately moved off a few hundred yards. Our
warships then sent a score or more heavy shells over to the Turks.
When the Turks shell any of our positions we always reply but shell
some other of their positions, not the gun that is shelling us and this
always seems to shut them up. Quite a lot of shelling was done by
artillery on both sides at dusk. I intended getting a "dugout" made
for myself today, but did not get time. However, just at dusk I discovered a commenced tunnel – about 6 yards – right into the solid rock,
a few yards from our lines, and it was an ideal place, took up my
quarters with my orderly there. I find it was intended for some
machine gun work but was discontinued 3 weeks ago.

Weather: Fine & warm though a little cloudy – light clouds.

SUNDAY – DEC 12TH/ 15
Today is now over for it is 10 p.m. and am writing this in my tun-
nelled "dugout" prior to "turning in," for I have just finished my
work for today.

It has certainly not seemed a bit like Sunday for there has been an
extra ordinary amount of shelling today by both sides but particularly by the Turks. About noon today they shelled us for an hour & a
half and it constituted the biggest shelling they have ever given us
here. They must have sent over 300 shells and they all landed within
500 yards of where my men were working. They were fired at Hill
10 where we have a strong Battery, and as it happened I was within a
hundred yards of nearly all the shells, as this was the extreme left of
our allocation in the Second Line, being down there making a plan of
our portion. I continued my work, (for I was in the trench) until it
got too hot. Every time we heard a shell coming within 40 to 50

yards of us we had to lay down in the trench and squeeze close to the parapet. Some of these shells were very large and powerful and would throw a boxcart load of stones & earth into the air in all directions and fragments of iron from the bursted shells would fall everywhere. Several times as I was lying down small pieces of stone and lumps of earth fell on me and one shell burst – and it was a large H.E. – just about 6 yards in front of the Trench and sent earth and stones all over us, so I then discontinued my plans and went back to where my men were working and left the completion until after dinner. The Enemy used two new powerful batteries this morning for the first time – a German one and an Austrian one – and sent over a large number of 10" H.E.s. I do not think they did much damage.

I spent much of today making a plan of our 600 odd yards of Trench, with Barbed wire, machine gun emplacements, communication trenches, cookhouses, latrines, Headquarters, etc., etc., doing it all "to scale" and showing length in yards of all sections of Trench etc. The men continued the work commenced yesterday. I had Genl. Caley – Brigadier – & Brigade Major Wilson, along today to see how things were going and they were pleased. I received a wire about 10 mins ago from Brigade Hdqtrs that Dvn [Division] Genl DesLisle [de Lisle] would visit the lines tomorrow evg. 5.30 p.m. Capt. Alexander & his orderly came down this Evg. Capt A. has been unwell lately and has very sore feet, so he has come down to rest. They are also staying in my Trench.

Weather: Fine & warm

MONDAY, DEC 13TH/15

Nothing unusual occurred today. We continued our work as yesterday and made very satisfactory progress. It is now a certainty that we are evacuating, for stores are being shipped off all the time, also small lots of troops. It is rumoured that we are not evacuating the whole peninsula but only that portion of the Line left of Chocolate Hill retaining the Anzac–Cape Helles portion. We are making full preparations in the Second Line, in case we have to fight a Rearguard action. And of all the Brigades in this vicinity this very responsible duty and honour has been placed upon the 88th Brigade to which we have the honour of belonging. But we are all hoping and expecting to have no difficulty in getting off without an action at all.

At 5 p.m. I had all the men in my command placed at intervals along our 500 yds of Line in preparation for the G.O.C.'s [General Officer Commanding] visit, but he did not put in his appearance until 6.30 p.m. There were nearly a dozen Staff Officers including the

G.O.C., the Brigadier (Genl Caley) & Br. Major (Wilson) They expressed themselves as pleased with everything.
Weather: Fine and Warm

TUESDAY – DEC 14TH/15
Everything today was quite normal – there was a little shelling by both sides but as far as the Turks shelling was concerned it was very ineffective. By the way, in that big shelling given us by the Turks on Sunday morning, our losses amounted to only one man killed and one man wounded. We heard today that there is a possibility of our Battalion going straight to the Beach from the Firing Line. I will then get special orders regarding my detachment.
Weather: Fine, but cloudy, and warm.

WEDNESDAY – DEC 15TH/15.
We built cookhouses for the various Companies today. After dinner 30 men unfit for the Firing Line (sore feet & c) were sent out to me and I sent in 30 of my best men to take their places. We heard that an attack by the Turks was expected tonight as the Turks had removed a large portion of their barbed wire last night. I therefore made arrangements to man my portion of the Second Line with my 50 men at the first sound of an attack.
This evening I received a note by messenger from the Staff-Captain of the Brigade to send all my ammunition (50,000 rounds) and all picks & shovels (about 200) other than what I absolutely needed, to the Beach tonight for evacuation, for which he would send me 5 mule carts.
Weather: This morning on getting up we found the wind had gone around to the north & it was quite cold. This was good for our evacuation for it would keep away the rain and gave us an offshore wind.
P.S. I sent one man to Hospital today with Dysentry.

THURSDAY – DEC 16TH/15
There was nothing untoward happened last night in the way of an attack. The men continued preparing the 2nd Line in case we should have to occupy it. After starting the men after dinner I went up to see the Colonel as there were one or two things I wanted to speak to him about and also wanted to see if he had any special orders for me. When I arrived at our Hdqtrs in the Firing Line the Col. said I was just the man he wanted to see. He said he wanted me for another matter & asked if I thought Capt. Alexander was well enough to take the Second Line over from me, so I said he was. He then asked me to pass things over to Capt. A. next morning and be back to Hdqtrs.

about Dinner Time tomorrow. He told me that that was a fine plan of our 600 yds of 2nd line I had sent him and that it gave him all information he desired to know. After discussing several other matters I was asked to have Tea before going back. After getting back to the 2nd Line, following out orders I had all the men's packs marked and shipped off to the Beach for evacuation. Tonight for Tea Capt. A. & I had jelly & Blanc Mange as did also our orderlies, N.C.O.'s & cooks. We had some also for supper.

It was made by Sergt. Bethune & the Cooks and was perfect and, as one will readily understand, quite a luxury & a treat. We got quite a large number of jelly powders and tins of Cornflour from the R.A.M.C. [Royal Army Medical Corps] Hospitals near us. We also got many other things such as Flour, Milk (ideal), Raisins, Currants, Jam Etc., Etc. We have had raisin and current dumplings and the men have had double rations since coming down and are naturally delighted.

Weather: Fine with wind in the same quarter but very much warmer and a perfect day and it seems a shame we are leaving it.

FRIDAY – DEC 17TH/15
I made preparations for returning to Headqtrs. Sent two more men to Hospt. today. About noon passed everything over to Capt. A. and started back for the Firing Line. At Brigade Hdqtrs. I met Capt. Wilson, C.R.E., and he gave me two men to carry up my things for I was really unable to do so having another two miles to go. A couple of months ago I was able to go any distance with it, but can now feel that although I feel perfect in everyway I have no reserve like which I have always had. Before leaving he made me stay for dinner. Arrived back around 2.30 p.m.

Weather: Fine & very warm. Wind changed to S.W. Heavy fall of rain last night.

SATURDAY – DEC 18TH/15
Spent all morning in the Firing Line, on watch.

Yesterday afternoon at a meeting held at our Hdqtrs. the Col. told me he was sending our first party in the evacuation scheme away tonight and that it would consist of 152 men & the quartermaster (Lt. Summers) and that he wanted me to take command of the party, which was to leave tonight Saturday. Now, this morning the C.O. was down to Brigade Hdqtrs to a special meeting and when he returned called a meeting of the company Commanders – I was sent for also – he said the previous plan had been changed and that now only 100 men would go in tonight's party. I was not to go, but Capt. O'Brien would go in charge and the Quartermaster and four other officers. Instead, I was to go down to the 2nd Line tomorrow morning at 4.45 and take up a position there with 20 men, 8 machine gunners & 1 machine gun beside my orderly – a total of 30.

Capt. O'Brien & his party left the Firing Line at 6.45 p.m. tonight & would be well on their way, presumably to Mudros, by midnight.

Weather: Fine and warm but chilly at night.

SUNDAY – DEC 19TH/15
With my party numbering 30 and a machine gun, I left the Firing Line at 4.45 a.m. and reached the position that I was going to take up in the 2nd Line about 6.45 a.m.

We had a very uninteresting day beyond the fact that half a dozen very heavy shells landed dangerously near us, in fact one showered stones & earth all over us. In the afternoon for about three hours some very heavy bombarding was done by our navy in the vicinity of Acha [Achi] Baba. For at least the first two hours the shelling or rather the reports of the exploding shells, were incessant and continuous, there must have been a great many thousand rounds expended. It was certainly the heaviest bombardment we have yet heard; we could see very little of it as the atmosphere was very smokey. We presume the object was to attract the attention of the Turks to that quarter & thus facilitate evacuation of Suvla portion. At 6.30 p.m. I moved up to Brigade Hdqtrs on the top of the 2nd Line arriving there and reporting to the Brigadier at 7 p.m. He told me I would move to the Beach about 11 p.m. Several detachments from our Regt. left during the day, about a hundred at 5.45 a.m.; a couple of parties after dusk, and another party of 20 with mine. The remainder – about 20 are leaving the Firing Line at 1.15 a.m. – 2 hrs after our departure from 2nd Line. Lt. H. Rendell is in charge of this final party and with him is the Bombing Officer[8] Lt. Clift.

We arrived at Beach at 11.45 p.m., but they were not ready for us to leave yet.

Weather: Fine but slightly chilly.

MONDAY – DEC 20TH/15.
My party went on board the Isle of Man paddle boat "*Barry*" at 12:15 a.m. and left for – we knew not where at 1 a.m. We arrived at the Island of Imbros at 2.15 a.m. but by the time we got inside of the two lines of torpedoe nets, reported etc., it was 2.45 a.m. We then had to go ashore on lighters, etc., and I landed at 3.30 a.m. We then had to walk about 3 miles to our allotted area, where we found many other Regiments' or Battalions' details. Everyone was then

supplied with a good supper. Just as we were about to turn in, the Brigade Signalling Officer & I noticed a big flare in the sky over Suvla Bay, so we went up on a hill to see it, for it was the stores on the Beach which had been saturated with oil and which had now been set alight. It was certainly a big fire and lit up the sky all around. We watched it until we got tired and then went down and tried, without much difficulty to get a couple hours sleep, at 6 a.m. After Breakfast the C.O. told us to act as O.C. "B". Coy. as there was no B. Co. Officer here they having gone in the first detail which went to Mudros as far as we know. The valley that we are now in presents a very busy scene for there must be several thousand men here. Aeroplanes have been soaring around us this a.m. – our own, of course – and they have been very low down. The past day or two they have been continually over the firing lines – sometimes as many as four at a time.

Last night one was around all the time from dusk until the time we left, and we were saying that this would worry the Turks a lot and make them think lots of things. This morning was spent getting the Battalions and Companies together. About 1 p.m. Rendell and Clift with their party arrived. Everything even up to the last went off without a hitch. The Turks were apparently under the same impression they have been for some time, viz: that we were preparing to attack them, for they had been continually strengthening, improving and renewing their barbed wire, etc. I would certainly like to be overhead in an aeroplane when the Turks find out that they are opposite empty trenches. Then when they begin to move forward they will have all kinds of plots to contend with, for the R.E. have various kinds of wires laid, such as "trip-wires" and those which will explode when one walks on them, by a falling box etc. Then in many "Dug-outs" wires have been laid attached by a wire to a table-leg which will be exploded by a movement of the table, etc. The Turks will have a great time "looting" but they will get very little, for we have removed most things, including nearly all the guns, and damaged and burnt what little is left. This evening we were allotted to our particular area and on this there [were] some 23 tents which just managed to hold us. We had rifle inspection and this was all we did today. A German (or Turkish) Aeroplane came over this evening and flew around for about half an hour. Our Anti-Aircraft guns got to work and fired about 50 rounds at her; they did not do any damage but succeeded in driving her off.

Weather: Dull & Warm

TUESDAY – DEC 21ST/15
We had nothing to do again today except to get a list of our men's deficiencies.

Weather: It rained very heavily with light thunder & lightening for a couple of hours, fortunately we were in tents, so kept fairly dry.

WEDNESDAY – DEC 22ND/15
There were rumours this a.m. that we were moving, some rumours said to Mudros & some to Cape Helles. However, at noon we got orders to be ready to leave Camp at 1.15 p.m. at which time we left, arriving at Beach at 2 p.m. We then went on board a lighter and were taken off to a small steamer – the "Redbreast" – getting on board at 2.45 p.m.

In about an hour they had the various portions of units of the 88th Brigade on board to the number of about 1300. Even now we did not know whether it was to Helles or Mudros we were bound, but rumours seemed to point to Mudros. The "Redbreast" was a steamer which plied between Mudros and Imbros, and further had not yet been to Cape Helles. Then the men who brought the lighter out said the steamer was bound for Mudros. However, we had nothing to do but to "wait & see". We were told that there would be Tea for the Officers at 4 p.m. We had Salmon, English Corned Beef, Bread & Butter, Marmalade & Tea, and all enjoyed it immensely seeing we were able to sit to a table with a cloth on it (which had been clean) once more.

After Tea all sat around the Lounge and smoked for a while. After about an hour we commenced singing and had a great old time until about 9.00 p.m. when the anchor was taken up and preparations for starting were made. By this time we had heard almost definitely that we were going to Helles. At 9.10 we started, and as we would be an hour and a half going across lay down for a nap. All the staterooms had been engaged ahead so we heard by senior officers of the other Battalions and our Colonel was the only one who had one in ours. So we tried to sleep in all sorts of places – some in the lounge, others in the chairs in the saloon, and two others and myself slept very peaceably, on the Saloon Table. I got asleep almost immediately, but suddenly woke up, and the steamer just then stopped. I thought she had stopped in going out thro' the mines at the harbour entrance but was surprised when Capt. March came & said we had arrived at Cape Helles, for I thought I had been asleep less than ten mins. It was now 10.40 – just the hour & a half. We had to go ashore on lighters in two or three parties, and by midnight were all ashore, except 3 Officers & about 30 men, who were kept back for the purpose of landing some stores – Jim was in this crowd. When the main crowd had all landed, we formed up and started off for our lines with Col. Hadow in charge. We have about 300 men including the other 30.

Weather: Fine & cool at night.

THURSDAY – DEC 23RD/15

Having no guide it took us a long time to find our intended lines We could not find Rowsell who had come up with a few men who had landed in the first lighter, so we got somewhere near where our place should be and finding a lot of empty dug-outs put the men there for the night. I think it must have been 4 a.m. when we, the officers, turned in. After Breakfast we found that we were practically in our own lines and Rowsell & his party were not far away. We spent today in settling the men away.

Weather: Fine & a little cold at night.

FRIDAY – DEC 24TH/15

We had nothing to do this a.m. except one Inspection.

Half of our officers went up to the Firing Line this a.m. to look over the 29th Divs. portion, the other half were to go in the afternoon. At the time for the departure of the Second party – 2.30 p.m. the first party had not returned so we waited for a while. At 3 p.m. three of us – Butler, Patterson and I – started off. We got nearly to the Firing Line, to 87th Brigade Hqtrs, and were told that it was useless to go any further as this was a days job. We got back at 5 p.m. and found that the first party had not been long back, having taken practically all day. We put our turn off for another day.

Weather: Very fine all day, but slightly cloudy and cool at night.

SATURDAY – DEC 25TH/15

Christmas Day

Although Xmas Day it has been my busiest day since arriving at Cape Helles or rather I should say – I have done more today than any day since our arrival. At 1 p.m. I received a visit from the Adjutant who said that a party of 80 men had to go to 58th Bde Hqtrs. and bring 400 picks & shovels to the Hants, Essex & Worc. Regts. which would take 2 to 3 hrs. and that the Col. told him to tell me to go. Patterson took the party down and back & I had to look after the work and distribution, getting receipts for the tools from the various Regts. We finished work and returned to our "Digs" at 4 a.m.

We did nothing today but loaf and sit around. We had fairly good feeding, for we had succeeded in making a few purchases of chocolate, figs, nuts, tangerines, etc., from a very remote and very small

canteen which was very soon sold out. Nearly all the very few Canteens including the one on the main beach were sold out, having been
unable to get goods in lately owing to the Evacuation of Sulva. Quite
a lot of shelling was done this afternoon almost as much as last Sunday. The Turks commenced it but they certainly got much more back
than they gave as they always do.

At 7 p.m. I had to take these same tools brought over this morning, back to the 86th Dump, having first to collect them from the
other Regts. This took us again another 3 hrs – I was the only officer
this time. When I got back I found all the officers in the C.O.'s
Dugout having been invited by him to spend a few hours, smoking,
chatting & singing. We had an enjoyable time till midnight.

Weather: Very fine all day & night.

SUNDAY – DEC 26TH/15

At 9.00. Tait, Patterson & I went on a reconnoitring trip. We were
asked by the Brigade to send three officers to find the shortest way to
get to various parts of the Firing Line – viz: that held by our Divn. –
29th – that by the 42nd on our left and the 52nd on our right, so
that in the case of emergency we could get anywhere without trouble.
This occupied all morning.

There was nothing doing this p.m. At 5.30 p.m. our Batn. had to
send all our available men to do some trench digging for concealed
wire entanglement. Capt. March with Lt. Bartlett & Capt. Rowsell
with Lt. Rendell went with 150 men.

Weather: Fine with fair temp.

MONDAY – DEC 27TH/15

Today was an absolutely blank one as far as our Battalion was concerned. We had expected the arrival of the balance of our Batn.
sometime last night, but they did not turn up.

This afternoon a wire was received from the Brigade Hdqtrs. saying that the Nfld Batn. would not supply the fatigue party as last
night, but that the Hants, Worc & Essex would each supply 50 men
& 1 officer, besides their 150 & 3 officers so as to have the same
number of men working on our Brigade's job as last night. It seemed
that our Regt. did twice as much work as the other Regts. so the staff
decided on the foregoing for to-night. Needless to say, our boys were
exceedingly pleased, and we also know that the other Regts. must
have been very displeased. Our C.O. was delighted that the superior
work of our lads had been recognized.

Weather: – Early this a.m. about 3 o.'c. it rained very heavily with
thunder & lightening for about 3 hours, but am glad to say that all

of us kept practically dry for we had tarpaulins over the greater part
of nearly all dugouts. Fine during the day and tonight – the night
indeed was perfect. Wrote first letters today for five weeks. Home –
Aunt Laura – & Etc.

EXTRACT FROM LETTER DATED 27 DECEMBER 1915
During the past five or six weeks many changes have been accomplished. When one looks back over the various doings, it looks more
like five or six months.

We have been told that there are forty thousand bags of mail at
Mudros Bay, Lemnos, for the Peninsula troops. A short time ago they
had to make drains to drain water from the bags. They are sure to be
in a fine condition.

We expect now that we shall be off the Peninsula in a fortnight.

TUESDAY – DEC 28TH/15
This morning we had to supply 200 men and 7 officers for road-
repairing in Gully Ravine. We left about 8 a.m. in parties of 40 men
at 5 mins or so intervals. I was in charge of a party. We took our
lunch with us and did not return to Camp until 5 p.m. I had lunch
today with the R.E. of the 42nd Div. by invitation of one their officers.

Weather: Very fine & warm.

WEDNESDAY – DEC 29TH/15
Today we continued work commenced yesterday, leaving our lines at
7.45 having risen at 6 a.m. There was quite a lot of shelling done
over our part of the Gully today by the Turks, but they succeeded in
getting only one of ours, Moakler,[9] though we saw two or three others being carried out on stretchers. A mule was killed by shrapnel just
two or three yards from where I was working with my party of 40
men. The Turks were using a new kind of shell sometimes today and
which they have used once or twice lately: this shell fires both shrapnel and H.E. It is an ordinary shrapnel shell which bursts in the air,
throwing out besides the shrapnel a smaller shell which explodes
when it strikes the ground.

We were all working in long rubber waders reaching to the top of
our legs, for the mud was knee deep and more nearly everywhere. We
were endeavouring to make a large drain throughout this ravine for
the recent rains had destroyed most of the roads in the ravine and
nearly all this water was still in various parts of the ravine. We also
had many bridges to make. We did not get back until 8.30 p.m.

Weather: Perfectly summerlike.

THURSDAY – DEC 30TH/15

Having received orders from the Brigade last night we [went] out to W. Beach this morning where we were to take up our quarters for we were now to do some fatigue work, as the Greek Corps had refused to work on account of so many shells coming to the beach lately; they were beginning to lose quite a lot of men and were therefore afraid to work there.

There was general dissatisfaction over this for we considered that they were thus making "Navvies"[10] of us. However, we must "Stick it".

We all got out before dinner but unfortunately the Turks shelled us very heavily. We went out in parties of 20 at five minutes interval, but after a while the Turks saw us with the result the shells were soon falling all about us. We then divided our parties of four at 100 yds interval. One shell, though it landed 30 yds away got the whole of one of my parties of four. My orderly, Thos.Cook,[11] was one of them, he was wounded in the left leg & left arm, – our mess cook was another, he got it in the stomach, – another got a leg wound, and another a splinter in the heel. I went to the Hospital this evening to see them and they were all happy and feeling no pain, except poor Geo. Simms,[12] our mess cook, who died about 3 o.'c. Considering the circumstances and conditions, everyone thinks we were very fortunate in escaping so lightly, for many of us had some very close shaves – one just went over my head and burst less than 20 yds in front, and several burst almost as near.

This morning about 7 o'clock an enemy aeroplane was over our lines and dropped a half a dozen bombs. One dropped in the B Coy's. line but did no damage beyond blowing away some steps leading down to the dugouts, the remainder dropped clear of us except one which dropped on the parapet of our dugout but failed to explode, otherwise we do not know what the result might have been. Another one also, dropping further away, failed to explode. The explosion from these bombs was terrific & lot[s] of condensation which had accumulated on the inside of the Tarpaulin over our dugout was literally splashed down onto our faces, which alone was sufficient to wake us; however, this combined with the noise woke us very suddenly and rudely. We almost thought a shell had landed in our dugout and were surprised to find "ourselves alive".

It took the afternoon to get things straightened away.

We also supplied one or two small fatigues.

About a dozen shells have come very near us – no damage.

Weather: Summerlike.

FRIDAY – DEC 31ST/15
This morning we commenced work on various jobs in preparation for the Evacuation of Cape Helles, or in other words the remaining portion of the Gallipoli Peninsula now held by us. It looks as if at least the 88th Brigade (our Brigade) of the 29th Divn. will be in the final stages of this evacuation as well as Sulva Bay – which everyone says is a great compliment and honour. Am afraid this one will not be as successful as Sulva for conditions are not nearly as favourable here, however we hope for the best.

Some of our men were building piers, others quarrying rock, others carting the stones by means of trucks on rails, to the piers and others doing fatigue work of various sorts. There were periodic spells of shelling and whilst on we had to suspend work. It was surprising how near they would come without doing damage.

It is the intention to be off the Peninsula within a week, in fact we must for the stormy weather is likely to begin any day.

Weather: fine and slightly cool.

SATURDAY – JANY. 1ST/16
We continued work as yesterday and had the same inconvenience regarding the Turkish shells. The one we fear most is a 5.9 Naval gun fired from Asia Minor at good yards, as is a gun from the "Goeben."[13] Being on the Beach with sloping high cliffs on all sides and as the shells come over the cliffs on the left side we do not hear it until it is quite upon us and have not time even to wink an eyelid, so have no time to seek cover. However as the Turks send generally four or five we have time to get under cover before the second comes. The Turks can also enfillade our Beaches from the other side so they become very hot sometimes. This Asiatic gun is known by the men as "Asiatic Annie"[14]; there is also another known as the "Louise Loue."

Weather: Fine, very windy and fairly cold. It appears as if a storm is coming, but we trust not, for it would greatly interfere with our evacuation plans.

SUNDAY – JANY. 2/16
Continued work as yesterday, and pleased to say we have had no casualties as yet, though sometimes it is really marvellous how we escape.

Many troops, guns and stores are being sent off every night. We hear that 25,000 men went last night and some of our men saw lots of guns going off.

Everything is going along satisfactorily so far.

"Asiatic Annie" killed a Major M.L.O. [Military Landing Officer] and wounded his Lieut today. They were in their dugout.

Weather: As yesterday.

MONDAY – JANY. 3RD/16

Continued work as yesterday. Quite a lot of shelling again today, but no one of our crowd was wounded. We are progressing very favourably with the breakwater & pier.

Troops, vehicles, mules ammunition, etc., still continue being evacuated.

Weather: Very windy last night, but this morning the sea was as smooth as a millpond and continued so all day, so a push will be made to get things off in a hurry tonight.

TUESDAY – JANY. 4TH/16

Same work as above. The Turks did practically no shelling today, that is, as far as our beaches are concerned. It makes us wonder, for we feel suspicious as to what they mean or are up to.

This morning a fire occurred in "D" Coy's lines, burning out some 7 or 8 shelters belonging to 16 Platoon. Some of the men's clothing etc was burnt, but no serious damage resulted; no one knows how it happened as all the men were out working.

At 7 p.m. I sent over for Jim and was surprised to learn that he had gone off to Mudros with some other sick men in charge of Lt. Knight at 6 p.m. I was not only disappointed, for I had some clothing, etc for him which I had got from the ordnance stores, but I was vexed that he had not come to see me; he was going off tonight and would, of course, get off in safety, but I do not know when I am going off and may be unfortunate enough not to get off at all, seeing that we are working under shell fire all the time. Narrow escapes are so very common now that we pay practically no great attention to it. Also the nearer my turn to go off comes to the last night of evacuation , the greater the danger. However we all hope for the best though this evacuation will be a much more serious affair than Sulva Bay. Jim has not been well since that storm of Nov 26th and for two weeks more has been suffering from piles and now has a touch of Jaundice.

Of course it was the proper and only thing for Jim to do, to go off with tonight's party for he has been off duty for nearly a fortnight, but I would have liked to have seen him before he went. I was off working all day, returning about 5.15 p.m. and had tea immediately and forgot that our sick were being sent off tonight.

Weather: Fine, warm & calm.

WEDNESDAY – JANY 5TH/16

Last night the wind came up again and we were afraid it might have done some damage to our pier, but as the day wore on it went down and succeeded in doing no damage to any of our piers.

Today a shell came over the cliff and hit the ground three yards from our wooden hut. Knight, Bartlett[15] & I immediately came out and did not return for awhile. Though it shook our house and made a hole in the ground 2 feet deep and 4 wide it did no damage.

The Turks shelled the beaches a lot again today.

Weather: Fine, a little cold and windy early in the day.

THURSDAY – JANY. 6TH/16

Work as usual today. At lunch time the C.O. asked to see all the officers. Orders as to the embarking of our Batn. Various parties were also told off for various duties etc., for instance: one party was to go to X Beach for fatigue work, another to V Beach for fatigue work, another to W Beach for fatigue work.[16] I had to take 24 men to the P.M.L.O. [Principal Military Landing Officer] – General O'Dowda[18] – where they were to act as guides etc. We were to report at 8 o.'c. tomorrow morning and would not be leaving the peninsula until the last thing on the last night which would be Sunday night. However, after Tea the C.O. called another meeting of the officers as he had received fresh orders, the original plans having been changed. Originally, embarkations were to take place tonight, Thursday night, Friday night, Saturday night and Sunday night, whereas the new plan practically combined Friday and Saturday nights, so that the three final nights of Evacuation would be Thursday, Friday and Saturday nights.

I was still on the same job so would be here until the last. This is just what I was hoping for – I would like to see the very end of it. For the last night there would be Lt. [John] Clift.[17] Lt. Taylor & myself with 102 men. About 6 officers, including the C.O., the Doctor & the Quartermaster with about 150 men went off tonight.

Weather: Fine, cool & very little wind.

FRIDAY – JANY. 7TH/16

At 8 a.m. I reported to General O'Dowda with my Sergt, Corpl. & 22 Privates. The General, with his Major (Yates[19]) and his Captain then took us to the two "forming up" places and we allotted 10 men & the Sergt to "B" forming up place, and 4 men & the Corpl. to "C" forming up place. They were then all told exactly what they were to do. 3 men of the "B" party & 1 man of the "C" party had to stand at gaps in the barbed wire entanglement so that when troops came along they would give them the direction they were to go. The remaining men of these two parties were to guide the troops from the forming up places to the pier that they were to embark from. -There were four piers at our (W) Beach.

The remaining 8 men of my party had to act as orderlies at the P.M.L.O. Office.

The men, under the Sergeant, then spent an hour in becoming acquainted with their surroundings and the various routes to the piers. There was now nothing else for them to do until 5 p.m. tomorrow, so they would have today, tonight & tomorrow off.[20] This means that we will be very nearly the last off the Peninsula, and is just what I have been hoping & wishing for. I spent the day in packing up and *"resting"*.

Nearly all afternoon there was tremendous Artillery bombardment by both sides on the left of the line at the head of Gully Ravine.[21]

I went to the top of the cliff just beyond our present location to have a look at it and to see if there was anything serious on, but all I could see was smoke and continuous flashes.

Three of four of our battle ships took a very active part in it also, but from where I was one could not tell just how things were going, still, it had every appearance of an attack. If the Turks have attacked it is a good thing that they have done so today instead of tomorrow for the most of our artillery is going off tonight.

Whilst this was on, the enemy sent a few shells to the beaches. A few came to us as well; one went right into the "B" Coy's. Cookhouse. Several of us immediately went over and found a man named Moland [Mouland][22] very seriously wounded. He was wounded in the side of the face, the left arm, the right hand had several fingers almost blown away and there was a wound in the right leg. The wounds were temporarily dressed and he was immediately sent to the pier to go on board the Hospital ship. Whilst his wounds were being dressed we found another man under the debris but he was dead, his name was Morris[23] and he was very much cut up. Just as we were getting him out we heard another shell coming, and I must admit that I was more frightened at the coming of this shell than I have yet been. Having just seen the extensive damage done by the first shell, for everything was levelled to the ground and coal and burnt wood dust was covered over everything, and seeing the heartrending condition of Moland, made me feel and think what an awful & cruel thing war is, and then hearing another shell coming screeching through the air directly for us produced a very queer sensation indeed.

I cannot understand how this second shell got no one for it

landed only a matter of a few yards from half a dozen of us who were tumbling for the merest bit of shelter.

It is very strange, indeed, what trivial bits of shelter we make for under such trying moments. I have seen men, and have done so myself, dodge behind a small bush not larger than a gooseberry tree, being even worse than the proverbial ostrich, for such a shelter would be no protection whatever, but under such a trying ordeal, one feels a great sense of protection and safety behind the least little cover.

Whilst the bombardment was on an enemy Aeroplane came over and dropped bombs everywhere in the vicinity of "W" Beach, but we do not know what damage was done. About 7 o.'c tonight 4 men came & reported to us that they were carrying kits out from the firing line in the vicinity of Gully Ravine at the time of the shelling this afternoon, and had to leave the kits for shelter & were unable to get back again, so thought the best they could do was to come along the beach and report to someone. They said the Turks had attacked our line and had taken our Firing Line in the vicinity of the head of Gully Ravine, and that they were, to use their words "cutting up our men something awful", also that some 50 or 60 men from our Firing Line came running past them, having been driven out by the Turks and were running back to save their lives, Etc., Etc. and that they did not know what was going to be the end of it all.

We soon saw that these men had simply turned cowards and were nearly frightened out of their lives. Seeing they had "cold feet" we sent them to the Hdqtrs of the 8th Army Corps nearby. Tonight when I was up to "W" Beach I found that there was not the least truth in their statements, except that the Turks had attacked and that our men, seeing them come up over their parapet, simply waited for them, and just before they reached our barbed wire they let them have a withering fire which wiped them nearly all out. Have also heard that the Turks got by far the worse of the afternoon's bombardment and must have had very heavy casualties.

At midnight I was awakened by Mansfield[24] of "C" Coy. who had gone off with a party of our Batn. about 7 or 8 o.'c and as they had a walk of about three miles and as he had a sore knee, was unable to walk it, so dropped out and came back without permission. And at 1.30 a.m. I was awakened by 3 men more who had been with one of our working parties on "W" Beach and were detained on a lighter, storing away goods, longer than the rest of the party, and when they got ashore found that the rest of their party had been shipped off. I had therefore to add these men to my party of guides.

Weather: Fine, fairly calm & warm.

SATURDAY – JANY. 8TH/16.

At 7.30 this morning I took all my men over to W. Beach with their packs and placed the packs on board a lighter there for the purpose.

This morning I heard that Clift & Taylor with the men they had at V Beach went off last night. This means that apart from Lts. Rendell & Nunns, who are M.L.O.'s, I am the only officer from our Batn left here. I will have my 28 men and 32 others who are at W. Beach acting as orderlies, etc.

Having received word, I went at 12.30 p.m. to have a meeting of all officers in vicinity of W. Beach which was held in the P.L.M.O.'s Mess, for the finalizing of details for tonight, the last night of Evacuation.

At 4.30 p.m., I brought my men over placing 5 at "C" Forming Place under Capt Dunlop and 14 at "B" Forming Place under Major Yates[25] and took the remaining 9 with me to the P.L.M.O.'s Office.

The men were sent off in three batches – 7 o.'c – 11 o.'c & 2 o.'c. About midnight one of the magazines in which was stored Fire rockets, Flares, Fuzes [fuses], Small explosives, etc., was accidently set alight by a careless man with a candle. For a time things seemed serious, for a very large volume of flame was shooting out of the mouth or entrance to the magazine, and frequent explosions, & some of them very heavy, were occurring. There was a danger of it spreading to the other magazines which contained quantities of heavy explosives, shells, etc., and if such occurred it would not be safe on any part of "W" Beach. The danger appeared so great at one time that the officer in charge of the magazines advised us to clear out of the P.L.M.O. [P.M.L.O.]'s office. So we gathered up everything, but when moving out got word that the danger was past, though the fire continued to burn during the rest of the night, though with lessening severity.

Weather: Fine & practically calm all day but windy about 6 p.m. increasing to almost a gale inshore, by the time the last lighter left.

SUNDAY – JANY. 9TH/16

I spent my time assisting in the office and going to various Generals, Colonels, etc., with verbal messages, enquiries, etc. At 2 p.m. we received a telephone message from V. Beach that one of their two lighters for the last batch from Gully Ravine section had gone ashore and they were unable to get her off, so they asked for another lighter. As there was a heavy sea running word was sent for them to send their 160 remaining men along the shore to W. Beach. About 3 o.'c. a hundred men of the party under a colonel reached us. At 3.30 we were ready to go but there was no sign of the balance of the party which was in charge of Genl. Maud[e].[26] They had gone to the top of the cliff to come down by road as they had some stretchers etc. with kits and other things. They had evidently lost their way, so Genl. O'Dowda said they must be found, and asked who knew the roads and ground up over the hill, so I said I did. He asked me to have a look for him; I went up over the hill and shouted the General's name until I eventually found him and so soon hurried them on board the waiting lighter. This was a dangerous and fearsome undertaking when one considers the following. When I left it was 3.30 a.m. – The fuzes in the magazines were lighted at 3.15 timed for 45 mins. – one of the many fires to be, among the stocks of supplies etc., was already burning – the Turks had sent over a few shells during the night, though very few, but two at a time, and two were likely to come at any time – and again, our firing line had been empty since just before 12 o.c. (11.45 I heard) (4 hours) so one might possibly encounter a body of Turks.

However we got on board at just 5 mins. to 4, and within 5 mins. and before we had even untied from the wharf, the first magazine went off with a very heavy explosion. A great volume of flame shot hundreds of feet into the air, debris of all kinds went everywhere, and as we were only a hundred yards away from it some came our way. It did no damage beyond breaking one man's arm in three places. The Second explosion came when we were less than 50 yds off the wharf & flame, noise, and everything was greater though nothing reached us.

By now there were fires everywhere, and it was really a wonderful sight. Several took photos and I regret I had no films left for the roll which I had purposely kept for tonight & to last me to Alexandria. I found, this afternoon, would not fit into the camera. The sky was also well lit up by the fires at V. & X. beaches. Another lighter and a hospital boat went away from the pier when we did so we were really the last away from the Peninsula, for the last from V. Beach left about 3.15 a.m. Unfortunately for our sight-seeing desires, we were all ordered to get below, for it was very windy indeed, and the sea was beginning to wash over the lighter.

It was fortunate that we left when we did for the wind was increasing every minute and as it was dead in-shore there was a very heavy sea running. As it was we had great difficulty in getting away from the wharf & shore; in fact, we all thought we would go ashore, for after we were untied, the lighter drifted half way to the shore before we commenced to move seawards. What we were afraid of more than anything was that the fires and the explosions would draw Turkish shell fire. However, we started off about 4.15 a.m. Having received orders to go below we all do so but very reluctantly, for we

would lose the best of the sights. There were about 150 altogether
and a motley crew we were, ranging from Generals to Privates, and
as they had to batten down the hatches it was already becoming close
and smelly, so we knew what it was going to be like by the time we
reached Imbros, for we were going all the way by the lighter as it was
now too stormy to transfer to steamers as had been done up to the
present. This lighter, which is all that the name implies was handled
very playfully by the sea all the way over and instead of taking 2 hrs,
the time of the normal trip, we took 5 hrs. not reaching there until
after 9 a.m.

Fortunately, I was not in the least sick, though the trip was not by
any means pleasant, for many of the men were sick and were vomit-
ing promiscuously all over the place just where they were lying. I had
only 10 of my men aboard this lighter, the remainder having been
sent off during the night, principally to Mudros.

We sent about 25,000 troops off tonight – 16,000 being from W.
Beach.

Arrived Imbros about 9 a.m. Found 11 men more of our Batn., so
took charge of them. Got breakfast for the men and then got mine at
the Camp Commandant's. Then reported to the Camp Adjutant who
told me to get my men to put up tents for themselves and we would
get orders when we could go to Mudros. However, I took a walk to
the Pier and found that there were two Trawlers going to Mudros
this morning. I told the M.L.O. that I had 20 men to go & asked if I
could bring them down.

He said the parties were being detailed from the other end and
when we were told to go, to come down and walk aboard, but that
we would not be able to go by either of these. However, I went up
and brought my men down as I wanted to rejoin the Batn. as soon as
possible instead of wasting time fooling around Imbros, and the
M.L.O. told me to go aboard the first Trawler. Now the first Trawler
was somewhat smaller than the second which was lying outside of
her, so considering the heavy sea that was running I knew we would
have a better chance for our lives aboard the larger, so passed over
the small one to the large one. Of course, on these Trawlers everyone
has to remain on deck for there is no below deck accommodation,
and if I knew before-hand what the passage was to be like I certainly
would not have been so anxious to leave whilst the gale of wind was
blowing. The smaller lighter left about 12.30 p.m. and we left about
1 p.m. We passed the smaller about 2 miles out, and the sea was running very high then. The trip to Mudros should under normal conditions occupy about 3 hrs. but at the end of that time we still had
some of Imbros abreast of us with some 25 miles to go. The ship was

rolling something terribly, every roll one would think she was going over as her hand-rail would be continually going under water. Of course the decks were under water all the time and many, many times when she shipped water on one side and rolled the opposite way the water would be up past one's knees.

I was alright for about 3 hrs of it but then began to feel seasick and vomited a couple of times, but was not, I am glad to say, nearly as sick as I expected I would be, for it was easily the worst period I have ever been on the water. Nearly everyone was seasick and everyone was soaking wet.

After a very hard fight we reached Mudros about 11 p.m. but could not get in through the mine boom[27] so had to flash for assistance. A warship came & showed us the way in. Then after a lot of fooling around the large harbour eventually berthed alongside of the celebrated H.M.S. "*Aragon*" and went on board about 1 a.m.

The "*Aragon*" is the Mudros Hdqtrs of the Dardanelles Army staff and is fitted up very elaborately. The whole of the large lounge is fitted up into an up-to-date set of offices. She is of the Royal Mail Packet Line and must be nearly 15,000 tons. We were all glad to get off the Trawler so as to get our wet clothes off.

Weather: Cloudy & windy from morning till night. At night the sky cleared but the wind continued unabated.

MONDAY – JANY. 10TH/16

Got on board the "*Aragon*" at 1 a.m. and had a couple of cups of Tea & biscuits. Then got wet clothes off and was soon asleep between "*sheets*".

Arose at 8.30 and had a fairly good breakfast at 9 o.'c. Wrote a 16 page letter home to catch mail – this was the only letter I had time to write as the mail closed at noon. Shortly after lunch the ship's Adjutant informed me I could take my men aboard a lighter alongside which would take me to the SS "*Nestor*" for Alexandria. This departure from the "*Aragon*" was sooner than I had hoped for, as I was intending to wait until tomorrow before looking for orders. This morning I thought that the balance of our Batn was ashore under canvas somewhere here in Mudros, but now heard that they had all gone to Alexandria having gone with some 4,000 troops on Saturday night, which left it that I was the only officer of the Batn left here and had with me some 21 men.

I left the "*Aragon*" about 3 p.m. & arrived on board the "*Nestor*" about 4 p.m. The "*Nestor*" is a blue Funnel Boat of 14,000 odd tons, and is certainly a fine ship, though she does not come up to the "*Megantic*" on which we came out from England. When I got on

board the *"Nestor"* I found another 23 of our men, being the balance
of the party which left Cape Helles early on Saturday Night – there
was no officer with them so I added them to my party making mine
now 44 (including 2 Sergts, 1 Corpl. & 2 Lance Cpls). There were
only about 700 troops on board and we were to take some 2,500, so
will not be going until sometime tomorrow evening.

This is certainly a great harbour and can hold thousands of steamers. There must be fully 500 of all sizes here now, and an officer told
me he has seen it when things were in full swing at Gallipoli when
there must have been thousands of all sizes of steamers, trawlers,
warships large and small, which must have been a wonderful sight.

On shore the slopes of some of the hundreds of hills are covered
with tents, some being occupied by hospital people & some by
troops. There are over a dozen Hospital steamers here now as far as I
can see and at night when they are lit up with green lights & the
large red cross, they make a pretty sight.

Weather: Fine & warm and the wind has gone down.

P.S. I am in a two berth cabin with a Lieut. of the R.G.A. [Royal
Garrison Artillery] – Lt. Bidgood [NI].

TUESDAY – JANY. 11TH/16
I was surprised at noon today when the largest part of our Batn.
came on board. There was Capts. Alexander, March, & Rowsell, &
Lts. Tait, Knight, Bartlett, Patterson & Strong, and about 270 men.
Jim was amongst the number and was looking and feeling quite well.

Lt. Col. Burton was also with them and he was looking remarkably well and his hand was now quite well though one of his fingers
has been left slightly bent.[28] He said he was pleased to be amongst us
again, and I must say he certainly looked it.

This afternoon I received my first mail for about a month and a
half and got 2 bundles of papers from home 2 from A.[unt] Laura
and over a dozen letters. Needless to state, I had a very happy hour
or two.

Weather: Fine & warm.

CHAPTER SIX

Egypt and Europe Bound
(12 January 1916 – 3 April 1916)

The 29th Division, twelve battalions in all, arrived in Suez in mid-January 1916. It remained there until mid-March when, with eight other divisions, it was ordered to France and the Western Front. The brief respite in Egypt was used for unit training, interspersed with occasional games of football between units, a favourite pastime marked by lively competition as Steele notes in the diary.

For the Newfoundland Regiment the two months in Egypt provided a much needed break from the ravages of physical distress and disease experienced on the Gallipoli Peninsula. Even Steele acknowledged how much thinner he and his brother Jim were as a result of their prolonged sickness at Suvla. Although Suez was a mere 50 kilometres from the Turkish lines the regiment was not involved in the defence of the canal. Instead days were filled with parades, training periods, and long brutal route marches. The latter were particularly disliked especially in the intolerable heat of the Egyptian desert. The new commanding officer, Lieutenant-Colonel Hadow, a British officer with a background in imperial military service, quickly earned the ire of the regiment for his insistence on drilling his men under these conditions. One day, as Hadow headed off to a few days' leave, the men began to jeer. For Steele it was a source of great embarrassment, a "disgrace to the men themselves ... [and] an insult to the country from which they came."

Steele's sister carefully deleted reference to the incident from her typed version of the diary and letters but it remains in the handwritten version. Hadow, as Steele correctly indicated, was aware of the men's annoyance with him and pondered the outcome while on leave. He recounted in a later interview that he was troubled by the incident and returned early, determined to nip any dissension in the bud. He managed to re-establish himself with the regiment, working carefully with the senior NCOs to restore their confidence in him.[1] The only legacy of the incident was a popular ditty composed at the time that pokes fun at Hadow's harsh treatment of his "colonials."

I'm Hadow, some lad-o,
Just off the Staff
I command the Newfoundlanders
And they know it – not half;
I'll make them or break them,
I'll make the blighters sweat,
For I'm Hadow, some lad-o,
I'll be a General yet.[2]

A further incident concerned rumours that the men had not received their parcels from the Newfoundland Women's Patriotic Association. The WPA was formed in Newfoundland on the outbreak of war to send comforts to Newfoundlanders serving overseas. Spread out across the island the WPA organized the colony's women in a massive effort to knit socks, mitts, and mufflers, sew shirts, bake cakes and cookies, and raise funds for cigarettes, tobacco and special Christmas gifts.[3] It was a prodigious undertaking but of singular importance to the men of the regiment and a tangible link with home. Reports had filtered back to Newfoundland that some of the shipments had gone awry during the regiment's time at Suvla but, as Steele explained in one of his letters, military necessity made it difficult for priority to be given to the delivery of comforts.

Much of Steele's time at Suez was spent in training, learning the required skills of the infantry officer, notably the use and deployment of the machine gun. He made good use of his free time, socializing with fellow officers and taking long walks, his favourite form of exercise and a hark back to happier days in St. John's and the competitions that marked the summer months. Occasional references to home probably represent a yearning for an end to war and a return to family. Rumours abounded about the next posting. Some thought it would be Salonika; others, Palestine or Mesopotamia. Many hoped it would be "Blighty" or England but by March the consensus was Europe.[4]

As the Newfoundlanders headed away from Egypt towards the Mediterranean the details of grand strategy that would bring them to France and the Western Front were not known. Some, wishfully perhaps, hoped it would be England but Steele suspected Marseilles. He was correct. The trip to France was a welcome relief from the monotony of Suez and encouraged a good deal of carrying-on. Once the regiment arrived in France, and especially as it got closer to the Front, discipline became stricter, censorship tighter, and concerns about heavy action against the enemy more imminent. Steele consoled himself with thoughts of home and held to a sanguine belief that the war would be over soon.

WEDNESDAY – JANY. 12TH/16
Left Mudros for Alexandria at a quarter to eight this morning, with the promise of a fine trip, for it was quite a summers morning though a little windy.

At 9 a.m. Lt. Col. Forman [NI], R.F.A. [Royal Regiment of Field Artillery] who is O.C. Troopship called a meeting of the officers for the purpose of outlining plans of operations in the emergency of a submarine attack, for submarines are particularly active just now in the Aegean Sea & vicinity, in fact it was reported yesterday that two had been seen just outside Mudros.

At 10 o.c we had an Inspection Parade and at 10.30 a.m. a practice submarine parade was held the signal being given by blasts on the syrene. Everyone, of course, wore a life belt and the officers were told to wear their Revolvers & Ammunition.

Lt. Col. R. de H. Burton,[5] who, it has just been discovered, is the Senior Lt. Col. on board, assumed command of the Troopship today.

Weather: Fine & Warm

THURSDAY – JANY. 13TH/16
I am one of the three "officers of the watch" today, Tait & Patterson being the other two. My hours are from 4 to 8 p.m. this evening and 4 to 8 a.m., tomorrow morning.

We met six Trawlers today, which we learn are going up to a certain place on the coast of Asia Minor where a German Submarine Base had been discovered.

Weather: It has been quite windy all day today and about the middle of the afternoon it increased very much, [so] that by 6 p.m. there was a real hurricane blowing with occasional rain storms or squalls lasting about 15 or 20 mins which continued all night. The wind was the strongest I have ever experienced, and when the rain squalls came, it was very like our Nfld snow storms, only more severe than the worse I have seen. It was impossible to stand against it on the deck without holding on; further, at times one could not breathe when facing it.

The steamer is certainly a fine sea boat, nevertheless, she rolled quite a lot, particularly about 9 or 10 o.'c. p.m.

Several officers were absent from dinner, and during dinner several more had to get up and leave. After dinner several went to bed because, as they said, they were "not feeling quite up to the mark". My trip from Imbros to Mudros in the Trawler seasoned me for I had the pleasure of not being sick at all.

FRIDAY – JANY. 14TH/16
I went on watch at 4 a.m. and it was raining fast at the time, and continued until daylight 6.30 a.m. After this the wind died down somewhat, but this may have been due to the fact that we were drawing near the land.

We sighted the first land at 8.15 a.m. and arrived in port 10.30 a.m.

A couple of Staff Officers who were on board and 3 or 4 Cols. went ashore, for we had anchored in the harbour to await orders to berth. They were on board again for lunch. No one was to go ashore until the M.L.O. had come aboard with our Disembarking Orders, so we were all anxiously awaiting his arrival. He came aboard just before dinner about 6 o.'c. and we found that we were not to berth before tomorrow, so all officers were allowed ashore. I was anxious to send a telegram home owing to the fact that practically no mail had been sent home for nearly two months at least. I went with Patterson and Strong, and we found that owing to a super-abundance of first class fully paid matter going across the Atlantic by the Anglo-Amer Tel. Co.[6] that they refused to accept cross-Atlantic Deferred Rate Cables, and that the only rate was the regular one of 2/ – per word. As we had practically no money & I could not get any we could not manage it, so we went aboard again, having been away only 2 hrs, half of which was spent going and coming in a boat.

Weather: We were surprised to find it rather cooler than we expected. It was also rather damp & cloudy, they had had rain lately.

SATURDAY – JANY. 15TH/16
After Parade I spent the morning writing letters. After lunch the officers were allowed on shore until 7 p.m. but one per unit had to remain. I said I would stay. (For I had several more letters to write).

About 4 p.m. Cluny Macpherson[7] (now Surgeon Major) came on board, and as it was just about Tea time I asked him down for Tea.

By the way he & Lt. Donnelly (who left us nearly two months ago for hospital) were on board last night. I was speaking to the Doctor then. After we had Tea we went to the Smoking Room and chatted for about an hour. I had my correspondence with me & seeing my letters for post he offered to take them so I gave him about a dozen. After Dinner Officers were allowed ashore again. (I omitted to mention that we moved into a pier at lunch time). I went ashore to-night with Capt. March and when driving along a street we came across Capts Alexander & Rowsell & Lts. Tait, Donnelly & Patterson, so joined up with them. We went to the "Kursall", a variety and moving picture house. It was very poor so we left it & regret now we did not remain to see the pictures, for we have since heard that there were pictures of the Newfoundland Regt. shown. Some were Aldershot views and it is a very strange coincidence that these should be here just now. We then went & had a small supper and came home early for there is very little of interest in Alexandria.

Weather: Finer & warmer.

SUNDAY – JANY. 16TH/16
Some of the troops on board received orders to go last night at 9 o.'c. – others this morning at 7 o'.c. and we – that is, – the Newfoundland Regt. details – at 12.45 p.m. We left Alexandria at 1 p.m. exactly, and the weather was quite fine and warm. As we got further in the country it got much warmer, and we felt much more power in the sun. Lt. Col. Burton did not come with us, having gone ashore yesterday just after lunch, saying he did know whether he would be back or not.

We had an uneventful trip and enjoyed the ever charming country scenery until darkness came. About 7 or 8 o'c I and the others in our compartment went asleep. I awoke once or twice but soon dropped off again though we were all more or less uncomfortable.

At 2 a.m. I was awaked and told we had arrived at Suez, so went to see about falling in my men.

Weather: Fine & fairly warm at Alexandria and warmer according as we got into the country, but at night it was quite cool, almost cold.

MONDAY – JANY. 17TH/16
Arrived at Suez 2 a.m. and lay down in a Tent about 3 o'.c. Had nothing but my British warm[8] and found it extremely cold.

This morning was occupied in getting the men properly fixed off in Camp. There was nothing to do in the afternoon, so I walked down to Suez – about a mile – with Capt. March & Lt. Strong.

Weather: Fine and warm but quite cool at night. It becomes very warm by noon, but is not at all uncomfortable.

TUESDAY – JANY. 18TH/16
We commenced before Breakfast work this morning starting at 7 o.'c. and had 3/4 hr's walking etc.

At 9.30 we had Rifle and Ammunition Inspection, and at 11 a.m. Inspection of Kit & list of Deficiencies taken. We had the afternoon off.

After Dinner (7 pm) Capt. Alexander asked me if I would take a walk into Suez with him & as I had nothing to do I went for a walk.

Weather: Fine & warm and warm also at night.

WEDNESDAY – JANY. 19TH/16
We commenced today what was to be our regular daily work.

Company Parade 7.15 to 7.45 a.m. (principally smart walking)
Battalion Parade 9.15 a.m.
Nothing unusual today. Stayed in Camp all day.
Weather: fine & warm.

THURSDAY – JANY. 20TH/16
Parades as yesterday. The 9.15 a.m. Parade is the C.O.'s Parade and lasts until abt. 12.30 p.m. We march to a spot in the Dessert about 2 miles from our Camp and spend the morning at Rifle Exercises and Manuals, Marching Discipline and Formations. It gets very warm by midday and cools off again towards 5 o'c or 6 o.'c. It is very cold on the 7.15 Parade.

About 4 o.'c. Lt. Strong & I went to Suez then continued on to Fort Tewfik – a mile further. We did a little shopping, then had dinner at Hotel Sinai. After Dinner we strolled over to the club and found some "movies" were being shown, being shown every Thurs. night only.

Weather: – Fine & warm except early morning and late evening-as usual.

FRIDAY – JANY. 21ST/16
Parades as yesterday. This evening Capt March and I went to Suez and then to Port Tewfik. About 5.30 pm we went to the club and were pleasantly surprised to find a "Dance" in progress. It was given by the officers of a warship which is leaving here. We did not know there were so many white folk within miles of here. We did not dance for we had heavy "iron studded" boots on. It stopped at 7 pm (Dinner time).

Weather: – As yesterday

SATURDAY – JANY.22ND/16
Paraded as yesterday. It was my turn to stay in Camps to-day for
"D" C. – one officer per coy staying in each afternoon & night.

SUNDAY – JANY 23RD/16
When writing this date it has only now come to me that to-day is
Father's birthday.

At 9 a.m. I had to report to the brigade Major, who showed me a
new camp ground we were to move to, and with the six men I had
brought with me, had to peg off the ground.

At 4 p.m. the Battalion brought up their tents and pitched them. It
was a short distance only from our old grounds.

We are in fine camping ground being level sandy desert, and are
about a mile from Suez by the side of the Railway.

On the City side of us there are many fine looking Palm trees.
Camped near us is a Battalion of fine sturdy Ghurkhas, they look
very much like Japs; there is also a Battalion of Indians – Sikh besides
the 29th Divn.

Weather: Fine & warm, but chilly towards dark.

EXTRACT FROM LETTER DATED 23 JANUARY 1916
We are camped in a level part of the Sandy Desert near the Railway
about a mile from Suez, and are quite comfortable. We go to Suez
sometimes, but there is absolutely nothing to see or do there. We are
hoping to get a couple of days in Cairo for refitting, dentistry work
&c.

We are expecting to be sent to France early in April – to march on
to Berlin and end the War.

We have seen lately by St. John's papers that quite a sensation has
been created by reports received as to the care and treatment of 'Our
Boys' at the front. When one considers how many of our women-folk
have either husbands, sons, fathers or brothers in our Regiment and
the tremendous amount of money they have all expended for 'Our
Boys' it is not only a shame but a crime, for, on the whole, we have
been very well treated indeed, as anyone who looks further than the
end of their nose, or his own selfish comforts, will fully agree.

It is true we have not received very much of what our w.p.a. originally intended for us, but we should consider all that they have done
and 'take the will for the deed;' in fact, our w.p.a. performed the deed
as far as in their power lay. But we have received things from other
sources, and have been really well-treated by the Authorities. Of
course, there were exceptional periods when we could not get what
we required, and what we would like in the way of food, but this

was due to the weather which at periods made it impossible to land anything at Gallipoli for days, and, as far as food was concerned, we had to live on "Bully Beef and Biscuits," until the weather improved. Then for the month that operations re the Evacuation of Suvla, Anzac and Cape Helles were under execution, transports could not be spared nor could time be wasted in bringing more things to the Peninsula, for the transferring of immense amounts of war materials from the Peninsula to the nearby Island of Imbros, required a very large number of shipping of all kinds, as everyone will readily understand. From about the middle of November to the middle of January, practically everyone of us had no change of underwear; but this was absolutely unavoidable, owing to the extreme or rather very extreme conditions and happenings, for which no one was responsible, and any complaints that there were, came from a very small minority, for the majority looked at things in a very sensible manner. Personally, I think that everyone of us should be extremely thankful that we got off the Peninsula with our lives, and think nothing of anything we had to undergo or put up with whilst there. There were times when we did not get things we desired and when we would have liked, but we all know that such are the exigencies of war. One lot of things came in to-day, and we have heard of quite a lot more on the way.

However, mother, you need not bother about Jim and me, for we have enough of nearly everything now, and will have from now on. I have just received most of my Kit, some of which I had not seen since last October, so am now quite well off in that line.

Heaven knows our Newfoundland women have had quite enough anxiety regarding relatives without being freshly alarmed by false reports re hardships, &c., and they have worked more untiringly for us and the Empire and more loyally than any others in the world; and even if such reports were true, no mention should have been made of it.

MONDAY – JANY 24TH/16
Morning Parades as usual nothing unusual today.
    Weather: Fine & warm in morning but showery during afternoon, & cold at night.

TUESDAY – JANY. 25TH/16
We had our 7.15 Parade, but when the time arrived for the 9.15 Parade there was quite a Sandstorm on so it was called off. The Sandstorm was just like a small snowstorm. It was quite a small storm but too bad to go off, for one could easily lose one's way. About the middle of the morning two special Armourer Sergts sent

by the Brigade inspected our men's Rifles. All damaged parts were to be repaired or renewed. Stayed in Camp again to-day.

Weather: Sandstorm in morning. Fine & warm afternoon

WEDNESDAY – JANY. 26TH/16
Parades as usual to-day.

This afternoon a football practice match was held in preparation for a series of Brigade matches – I played.

This morning Lt. Donnelly & Lt. Fox with 57 N.C.O.s & men arrived to rejoin the Battalion – they were men who had been sent away from Suvla, sick.

Weather: Fine & warm, but cooler during afternoon owing to a breeze blowing.

THURSDAY – JANY 27TH/16
Nothing unusual today

Weather: Fine & warm – cold at night.

A series of Brigade football matches commenced today. Worcs [4th Bn Worcestershire Regiment] defeated Newfoundland 1 to 0.

FRIDAY – JANY 28TH/16
This morning we had an Inspection by our Brig. General – D.E. Caley and Major Genl DeLisle of the 29th was also present. We were put through the Manuals and "March Past" twice.

We acquitted ourselves as well as either of the four other Batns – Hants, Worcester, Essex & Royal Scots. Our Dr and our Qtrmaster, who were lookers-on, said the Royal Scots were best, which was due to the fact that instead of a Batn, they had about 150 men and they said we were easily the next best.

Weather: Fine & warm & cool at night.

SATURDAY – JANY 29TH/16
Parades as usual

Tonight we invited our [88th] Brigade Staff – Br. Gen. Cake, Br. Maj. Wilson[9] & Staff Capt Armstrong – to dinner.[10] We had quite an excellent dinner which included chicken and plum pudding. After dinner we spent a couple of hours singing various songs – well-known & otherwise & also sang favourite Regt Song composed by the Dr [Lt. Frew] assisted by the C.O. entitled "When this bloody war is over oh how happy we shall be".

The dinner was an entire success from beginning to end. The idea originated by someone saying we should have a "Christmas Night."

Weather: Fine & warm.

SUNDAY – JANY 30TH/16

Sunday being practically a free day with us I decided to walk to some
hills, which appeared to be about 4 or 5 miles away, and endeavour
to climb them. We obtained permission from the c.o. to be absent for
the day. I had asked the Dr. (Lt. Alexander Frew) to go with me and
we left Camp at 11 a.m. We took our lunch with us including some
cake, chocolate etc., of recent arrival and also took my Revolver and
Ammunition in case we came across any arrivals or "blood-thirsty"
arabs.

We reached the base of the hills at 1 p.m., but the hills themselves
were half an hours' walk further on, thus we took 2 ½ hrs to reach
the hills and we calculated that the hills were practically 10 miles
from our camp. We were surprised to find that instead of one big hill
with steep sides we were face to face with a succession of hills or
pinacles, each successive one being higher than the one in front of it,
some of which were absolutely unclimbable. We entered the hills by a
tremendous and in some places extremely beautiful Ravine. Its beauties, to be fitly described, require an abler pen then mine. There were
pinacles, gorges, ravines, caves, etc., full of artistic beauty. Fortunately I had brought my pocket Kodak with me and took some fine snaps
if they will only turn out O.K.[11]

However, we continued up the ravine and selected what we took to
be the easiest portion to climb and decided to wait for our lunch
until we had reached the top of the hill which we expected would be
about 2.30 p.m.

We climbed one hill then another which we thought would be the
top of the main hill or hills. About 2.30 p.m. we reached the top of
this hill and were surprised to behold another larger valley with a
higher hill on the other side.

As it would take another hour to reach the top, and as the coming
down would be more dangerous than the going up, we decided to
forego, for this time, any further attempt to reach the top, so sat
down to lunch. Further as it would be dark between 5 o'c & 6 o'c.,
we were afraid to run the risk of spending the night on the hills, for
the nights are very cold and we had nothing but a pair of shorts & 1
shirt on, with boots, socks & puttees.

About 3.30 p.m. we started the return journey and reached camp
at 10 mins to 7, being just in time for dinner (7 o'.c). We found that
the c.o. was just beginning to get anxious about us, and he told us
that he had just been speaking to the Adjutant about organizing a
search party to go look for us.

I decided I would go again next Sunday but the doctor said he
would not go again, for he had had enough of it. We reckoned we

had travelled over 25 miles. We both thoroughly enjoyed ourselves.

Weather: Very fine and very warm.

MONDAY – JANY 31ST/16

We had our usual 7.15 a.m. parade, but the 9.15 parade, instead of being a Battalion Parade was a Company Parade, each company going off independently. It will be thus until further orders.

Quite a lot of mail has been coming in lately and today 200 bags of letters and parcels came in .

Weather: Fine & warm

Football match: Royal Scots defeated Nfld 2/0.

TUESDAY – FEBRY. 1ST/16

Parades as last order. Another 65 bags of mail came in today.

During the past few days I received something like 46 letters & parcels, some being of October's date. Jim received about 34. However Jim & I have received over 80 packages between us.

Weather: Fine & warm.

WEDNESDAY – FEBRY. 2ND/16

Parades as last order. The C.O. said this a.m. that he is thinking of having a route march to the hills, tomorrow, and back again, in preparation for a later attempt to take the battalion to the top of the hills, if we can find an easy way of getting there. He said our battalion must be the first to get there.

Note: I omitted to state that Capt Raley, Lt. Taylor, Sergt. Major Power[12] arrived from Alexandria on Monday – Jany 31st about 10 a.m.

Weather: Fine & warm but the nights still continue cold, and tonight is colder than it has been for over a week.

THURSDAY – FEBRY. 3RD/16

No 7.15 a.m. parades but had breakfast at 7 o.c and went for a route march at 8 a.m. towards the hills.

Capt. March was unable to go owing to having rheumatism in his shoulder. He has had it for several days and it was worse today, so the Adjt. told me to take charge of "D" Coy being the next senior officer.

We arrived back at 1 p.m. About 24 men in the Battalion dropped out owing to chafings of feet and of legs, walk being too much, etc., etc.

Weather: Fine & warm – nights still cold.

FRIDAY – FEBRY. 4/16
Parades 7 a.m. and 9:15 a.m. I took charge of "D" Coy's 7:15 parade
but Capt. March the 9:15 parade. Company and Platoon Drill.
   Weather: As yesterday.

EXTRACT FROM LETTER DATED 4 FEBRUARY 1916
New Camp, Suez, Egypt.
We are now settled down here and are having fine warm weather,
though the nights are still cold.
   Your idea as to where we were spending Christmas was, unfortu-
nately, a little different from the reality. Of course, Harold
Mitchell's[13] misled you, but he left the Peninsula through sickness,
and you have, by this time, received letters telling how and where we
did spend Xmas.
   Yes, our Regiment has made a good showing in the matter of
M.C.'s [Military Cross] and D.C.M.'s [Distinguished Conduct Medal],
for we have two M.C.'s (won by Capt. Bernard and Lieut. Donnelly)
and two D.C.M.'s. (Won by Sergeant Greene and Private Hynes[14]).
Then Fitzgerald,[15] the Red Cross man, who was killed whilst attending a wounded man, after he had already been wounded, has been
recommended.
   Although we are here supposed to be resting, we get very little time
to ourselves, and are now getting plenty of afternoon work. Everyday
we march a few miles into the desert and spend a couple of hours of
more drilling, and in the afternoons we have to lecture our platoons.
   Have been in to Suez a couple of times, but it is not worth visiting,
for it is [a] small and very dirty place.

SATURDAY – FEBRY. 5/16
Parades as yesterday. Football match this afternoon. Essex [1 Bn
Essex Regiment] defeated Nfld. 2 to 0.
   Weather: fine and warm.

SUNDAY – FEBRY. 6/16
Lt. Strong and self started off at 8 a.m. to climb the hills which the
Doctor [Lt. Frew] and I failed to do last Sunday. [Typescript adds,
from a letter, We reached the top all right and built a six foot cairn of
stones and placed our Regimental Flag on top. We covered about
thirty miles.]

MONDAY – FEBRY. 7/16
Parades as usual or rather as last ordered.
   Nothing unusual occurred.

I commenced a Machine Gun Course[16] at Brigade Hd-quarters under B.M.G.O. [Brigade Machine Gun Officer] Capt. Windeler. There was an officer and an N.C.O. from each Batn. in the Brigade. The course will occupy about two weeks. We go for 2 hrs in the morning and 1 hour in the afternoon.

In the afternoon a team from the officers of our Batn. played the officers of the Essex – Essex won 3-0.

Weather: Fine and warm. Still cold at night.

TUESDAY – FEBRY. 8/16
I am now freed from all Company work whilst taking course. I attended Machine Gun Course. The Batn. started off at 8 a.m. on another Route March to the hills. They reached back about 1:30 having covered at least 15 miles.

About 4:30 p.m. the C.O. left for a few days holiday.

[Note: The rest of the entry for this day was omitted from the original typescript because of the references to ill feeling between the men and the C.O., Lieutenant-Colonel Hadow. This may have been because the typed manuscript was intended for publication and the incident would have reflected badly upon both Hadow and the regiment.]

The men had as much marching today as they wanted and over 70 fell out on the way. The men who fell out are to be given 1 ½ hrs marching each afternoon for a week as training; nearly all the men were in poor humour and expressed among themselves their opinion of what they considered the harsh treatment of the C.O. whom they blamed for the fast pace that was kept up over the hard going sand. So when they saw the C.O. going away and knowing he was going for a few days, in a foolish moment some of them commenced hooting with the result that it was taken up by others all around. Some of the officers had to go down and stop them, but it had already gone to a disgraceful degree, for it was not only heard by the C.O. but by the Batns. on our right and left, some of whom, it was apparent, recognized who was the object of their jeers. Of course it was stopped immediately, but we, the officers, were really expecting the C.O. not to go, but he went. This act was not only a disgrace to the men themselves but an insult to the country from which they came. Of course, doubtless some of the men joined in the jeering not knowing what it was about.

Weather: Fine and warm but still cool at night.

WEDNESDAY – FEBRY. 9/16
Company Parades 7:15 and 9:15 a.m. M.G. Course occupied my attention a.m. and p.m.

Weather: At 7. o'c it was for the first time since our arrival extremely foggy, one could scarcely see more than a hundred yards. After an hour it quickly cleared away, for the sun got to work. It was also very cold, they say around here that it is the coldest for 25 years. After breakfast it got warmer again.

THURSDAY – FEBRY. 10/16
Attended M.G. Course. Company Parades as last. Now that the C.O. Lt. Col. Hadow is away, Capt. C. Alexander is acting C.O.

This afternoon we played "Footer" [football] with the Hants [2 Bn Hampshire Regiment] and lost by 4 to 2. However, as will be noticed, we have played five games to date, and have today made our first score.

Weather: Fine and Warm. Night cool.

FRIDAY – FEBRY. 11/16
Parades – Company Route Marching. Self: M.G. Course.

The Nfld. team played the Field Amb[ulance]. and were defeated by 1-0.

Weather: As Yesterday.

SATURDAY – FEBRY. 12/16
The C.O. (Lt. Col. Hadow) returned last night.

The Batn. had Company Parades and I was on my M.G. Course.

Weather: Fine and warm and glad to say that the nights are getting warmer now.

SUNDAY – FEBRY. 13/16
This afternoon Donnelly and I took a walk to the Suez Canal.[17] We were told it was 7 or 8 miles away but found it to be only about 5 ½ miles, we were in shorts so walked quite fast, getting there in 55 mins. We spent about half an hour there and were in the Trenches our troops used twelve months ago. Walked it back for time and did it in 45 mins. Donnelly running by my side as pacer. I did this by way of practice as there are Brigade Sports taking place in a month's time and they want me to represent our Batn.

Weather: Fine and warm. Night warmer than usual.

EXTRACT FROM LETTER DATED 13 FEBRUARY 1916
We are both well and contentedly happy and are having great weather, for it is not too hot.

As my Machine Gun Course is still on, I am still very busy and have no spare time for letter writing, but Jim has told me he will have an enclosure, so I expect his letter will be more interesting than mine.

This is a great place for us to 'pick up' or regain our health. After going through what we did at Suvla, we are all more or less run down. Some of us had become very thin – I for one. Even Jim's fat face of 'pre-historic' days was gone and he was slightly thin, but is already getting it *all* back again. I expect that I shall also get more or less fat, even fatter than it was my custom to be. However, time will tell, so we must only do as Asquith[18] says: "Wait and see."

We do not know where they will be sending us next, whether to Salonica (for the Balkans), Mesopotamia, France or whether we shall get in conflict with the Turks here, for we are only six or seven miles from our front line and the enemy is known to be thirty miles away, but they may never come any further.

I suppose you are now going through Newfoundland's usual winter, and 'poor us' are having really good summer weather. I hope something wonderful will happen to enable us to share with you Newfoundland's summer of 1916.

MONDAY – FEBRY. 14/16
Batn. was out on Company Route Marches again. Self on M.G. Course.
Weather: Fine and warm.

TUESDAY – FEBRY. 15/16
Batn. was again on Company Parades and self M. Gunning. This morning three aeroplanes were out together – there is a new aerodrome just erected here for, I think, three machines. At 4 p.m. Capt. Alexander and I went down to Suez, both of us having shopping to do.

It is rumoured today that our brigade has received orders to reorganize our Machine Gun Section, according to the new Company Formation. No one has full particulars yet as to what this formation is but we will probably know in a day or two. It is also said that the M.G. Companies will go to France. Whether Reid will be going or myself remains to be seen.
Weather: Rain fell last night and it was a bit showery today.

WEDNESDAY – FEBRY. 16/16
Company Parades. Self on M.G. Course. Nothing unusual today.
Weather: Fine and warm.

THURSDAY – FEBRY 17/16
There was a Brigade Route March today. They left at 8 a.m. and

returned about noon. It was not nearly as hard as our own Battalion
Route Marches. I was still on my Course.
Weather: Fine and warm.

FRIDAY – FEBRY. 18/16
Company Parades as usual. I had my first exam this a.m. It was on
Gun Mounting and the Mechanism etc., of the gun. I did O.K. except
in one of the motions of the mounting in which I was 2 secs. over –
this was due to a pin being put in by No. 2. (who assists in the
mounting[19]) becoming jammed.
    In the afternoon another Exam was held this time on "stoppages".
I got full marks here.
Weather: Fine and warm.

SATURDAY – FEBRY. 19/16
Company Parades. Part of "A" Comp. went to the Baths[20] at Port
Tewfik (nearly 4 miles away). In the a.m. I had another exam -Theo-
ry, Tactics, Methods of Fire etc. Will not know result for a week. In
the exams of Friday I came top of the Officers and second in the
whole class of about 25 Officers, N.C.O.s and men. Sergt. Jupp[21] in
our Batn. was top.
    In the afternoon I went with Capt. Alexander to Tewfick [Tewfik]
baths with the balance of "A" Coy. and "B" Coy.
    The Essex had sports this p.m. but I was unable to see them.
Weather: Fine and warm.

SUNDAY – FEBRY. 20/16
Yesterday the C.O. asked me if I would go to the top of the hills with
him today. Capt. Alexander and Lt. Butler with a dozen men whom
he was training as scouts also were going. We all left at 4:15 a.m. but
just after leaving I found that my water bottle was leaking so had to
return to get another for the bottle would have been empty by the
time I reached the hills. I arranged to catch them up. After changing
my water bottle I left again at 4:35 a.m. and after going for a half an
hour could still see no sign of the others for it was only moonlight.
However I continued to the hills taking a zig zag course but reached
the hills without seeing any sign of them though I had my field glass-
es with me. I then decided to go to the top of the hills and trust to
meeting them on top. I reached the top at 8:25 but saw no sign of
them and did not expect to for an hour at the least. At 9:40 they
reached the top – and came up a different way than I did.
    We had intended staying for two or three hours but the wind was
now blowing very strong and it was intensely cold and as we were all

dressed very lightly. We hurried thro our lunch so as to start the descent immediately. When we were nearly finished lunch Capt. Alexander and a couple of men who had found it too much arrived, being half-an-hour after the others. We commenced the descent at 10:30 a.m. and took a round-about way to get down, taking our time – at one time we sat down for an hour – reaching home at 6:30 p.m. Capt. Alexander, who had been behind most of the way (catching up when we halted), arrived in ten minutes after us with a couple of the men and admitted they were "all in". Two of the men did not arrive until nearly 8:30 p.m. The c.o. is a very fine walker and did exceedingly well and felt fine though a little stiff.

Weather: Fine but windy and cool.

MONDAY – FEBRY. 21/16
The Machine Gun Course being over I went on the Company Parades.

Weather: Fine and warm.

TUESDAY – FEBRY. 22/16
Company Parades.

Nothing unusual today. Hants and Essex from our Brigade went into the firing line on the other side of the [Suez] Canal to a distance of about 25 miles.

Weather; Fine and warm.

WEDNESDAY – FEBRY. 23/16
We and the Royal Scots were "Duty Batn." today and as our Capt. had charge, we had very little to do for we had done all the work last time so gave the Scots a turn.

Weather: Fine and warm.

THURSDAY – FEBRY. 24/16
Today "C" and "D" Coys were having an attack. "D" Coy was leaving earlier than "C" to take up position in some ruins 2 ½ miles away which we were to defend against "C" Coy. We left at 8 o'c and about 11 o'c saw some of the enemies' scouts in the distance. Lt. Bartlett had charge of the Left Flank, Lt. Knight of the Centre and I the Right Flank. As it happened, they came up on my Flank which we had seen was the mostly likely place. I had an outpost out and had all my men well entrenched so we were quite capable of holding our position.

When "C" Coy approached, the c.o. called us all in and addressed the officers as is usual after such work and discussed various points

with them. He said that both the Attack and the Defense were well
carried out.
Weather: Fine and warm.

FRIDAY – FEBRY. 25/16
Today we had a Battalion Parade and the four Coy. performed an
Attack on the Ruins we were defending yesterday. "A", "B" and "D"
were in the firing line, detailing their own Supports and Local
Reserves, and "C" was General Reserve.
Half way thro' the proceedings "C" Coy was sent around to make
a Rear Attack. When the Attack was over the c.o. again compliment-
ed us and said there was very little fault to be found with the manner
in which it was carried out.
Weather: Fine and warm.

SATURDAY – FEBRY. 26/16
We were again Duty Battalion but did not have very much to do.
In the afternoon most of the officers attended the Sports [Meet]
being held by the Worcesters. There was a very fine programme and
was very interesting and events well contested.
Weather: Fine and warm.
In the whole M.G. Course I came top of the Officers and second
top of whole class.
P.S. After Dinner tonight we had a little boxing in the Officers
Mess. Capt. March and Capt. Windeler had a bout; – they had three
rounds. Then Lt. Strong and I had a couple of rounds. No one else
would put on the gloves, so proceedings were adjourned.

SUNDAY – FEBRY. 27/16
Nothing doing today besides the usual Church Services. In the
evening Capt. A[lexander], Capt. M[arch], and myself went to Tewfik
and had dinner at the "Sinai Hotel" then went to the Club.
Weather: Fine and warm.

MONDAY – FEBRY. 28/16
I commenced a Bombing Course this morning. Battalion was on
Company Training.
Weather: Fine and warm.

TUESDAY – FEBRY. 29/16
Self on Bombing Course. Battalion went on a Route March – starting
at 7 o'c and returned shortly after noon.
Weather: Fine and warm.

WEDNESDAY – MARCH 1/16
Self on Bombing Course. Battalion on Company work.
   Weather: Fine and warm.

THURSDAY – MARCH 2/16
Bombing course still on. Battalion on another Route March. Had
breakfast 6 a.m. Batn. left at 7 a.m. returning shortly after
noon.
   Weather: Fine and warm. Small sandstorm in the evening last[ing]
for a couple of hours.

FRIDAY – MARCH 3/16
Still on Bombing Course. Battalion at Battalion Training.
   Weather: Fine and Warm.

SATURDAY – MARCH 4/16
Last day of Bombing Course so am now a "Qualified Bomber". Battalion on Company work.
   Weather: Fine and warm with another small sandstorm in the
evening.

SUNDAY – MARCH 5/16
Another day of Practical Rest.
   In the afternoon Capt. A.[lexander] and I went to an Hospital
where there was one of our men who had been badly handled by a
"gyppy" [Egyptian]. He was getting on fine though at first it was
thought he would lose his eye. We then went on to Port Tewfik
where we had dinner, returning to Camp about 10 o'c.
   Weather: Fine and warm, but late in the evening the wind came up
very strong which gave us another small sandstorm which lasted for
two or three hours.
   I omitted to mention yesterday that the C.O. sent for 4 or 5 of the
2nd Lts. including myself and told us we had been promoted to 1st
or Full Lts. He asked me if I had heard of it before, I said I had not
so he said he could not understand why I had not for the promotion
was dated Oct. 15, though he had not heard of it until their just having received a telegram from the Governor.

MONDAY – MARCH 6/16
This morning after having our usual half hour Company Parade
before breakfast, we spent at Battalion Training.
   Weather: Fine and Warm.

EXTRACT FROM LETTER DATED 6 MARCH 1916
Egypt
We are still 'here' but in all probability my next letter will be from a different address. General opinion is stronger, that we are going to the Theatre of War to which we thought we were going just before we left England.

The weather here is still quite pleasant, though the flies are beginning to be a little troublesome.

Many thanks for the boxes you have sent and which will probably arrive in a few days. Whereas we are very thankful for your kindness in sending another lot of things, I must ask to please not bother sending us anything further, for we have not only all we want, but so much more that we shall not be in need for some time to come.

TUESDAY – MARCH 7/16
Today was another Battalion Day which we spent at musketry.
    Weather: Fine and warm.

WEDNESDAY – MARCH 8/16
Today was Brigade Day and we had a Brigade Ceremonial Parade. The whole Brigade marched past in column. Brigadier General Caley in Command; Major Gen. DeLisle – 29th Division acknowledged the Salute. We then marched back past the Saluting Base in line of Companies in close Column. The general opinion was that the Newfoundland Regt. excelled itself and acquitted itself creditably.
    Weather: Fine and warm.

THURSDAY – MARCH 9/16
This was another Brigade Day. We had another attack with which the General (Br. Genl. Caley) was very well pleased indeed.
    Weather: Fine and warm.

FRIDAY – MARCH 10/16
Today was a "Division" Day with the 29th Division less one Brigade (87th which has just left here – presumably for France). Performed a general attack. It was presumed that a badly defeated Eastern Force was retiring along the Cairo Rd. towards Suez, being followed closely by a victorious Western Force, and that the Advance Guard of the latter, which advance guard was to be the 29th Division, came up to the Rear Guard of the Eastern Force about 5 miles from Suez. The Rear Guard of the Eastern Force was entrenched and was covering the retirement of their main body which, aeroplanes reported, was

embarking at Tewfik. Aeroplanes also reported that the road from Suez to Tewfik was blocked with Infantry and Artillery. We were to suppose, also, that our Artillery was helping us – Artillery effects were shown by artificial means with dust and smoke. We, the Advance Guard of the Western Force, were to dislodge the enemy's Rear Guard, which consisted of one Brigade, so that when our main body came along, they would have no delay, but could continue on and attack the embarking troops. The 86th Bde. did the actual attacking, while we – the 88th Bde. acted as General Reserve. Besides Brigadier Generals of the Brigades, and Major Genl. de Lisle of the Division, General Williams [NI], the Corps Commander, was present; he was well pleased with the attack, but said it was not carried out exactly the way he would have done it. We left Camp about 7 a.m. – having got up at 5 o'c – and returned about 1 p.m.

Weather: Fine and warm.

SATURDAY – MARCH 11/16
Company Parades out on the Desert as usual.

Weather: Fine and warm.

SUNDAY – MARCH 12/16
I went to the C. of E. [Church of England] Brigade Service. We had the 29th Division Band there, and it made the singing go fine – it is quite a good band.

Weather: Fine and warm. The weather is fast getting warmer, and the flies more numerous and consequently more annoying.

MONDAY – MARCH 13/16
Company Extended Orders Drill. In the afternoon Capt. Alex., Capt. March and self went down to Suez and had dinner at the Hotel Belair and returned to Camp about 9:45 p.m. About 3 p.m. today we were suddenly given our orders to prepare for leaving and to be on board the H.M.T. *Alaunia* (Cunard Line – about 13,000 tons) at 1 p.m. tomorrow.[22] We do not know where we are going – the general opinion is that we are going to France, though some say, and *all hope*, we are going to "Blitey"[Blighty or England].

Weather: Fine and warm.

TUESDAY – MARCH 14/16
Reveille 5 a.m. Breakfast 6 a.m. "Struck Camp" about 7:30. Then cleaned up the whole grounds and put everything in order. At 11 a.m. we left for our late home and marched in full marching order, to Tewfik – about 4 miles. We were all on board by 1 p.m. Moved away

from the wharf at 4:15 p.m. and stayed in the stream a couple of
hours. The *Alaunia* is a fine ship, though not quite as good as the
Megantic. At 6:30 p.m. we started up the Canal, and up to bed time
had stopped for a couple of hours in two or three places. It was a
bright moonlight night, so we could enjoy a few very nice views – I
wish it were being seen in daytime.

Weather: Fine and very warm with cloudless sky – in fact all days
are alike here.

WEDNESDAY – MARCH 15/16
Arose at 6 a.m. – splendid morning and view of Canal interesting.
Arrived at Port Said ... – Took several snaps. At 2 p.m. – we were
then anchored about 50 yards off the wharf taking coal aboard – a
notice was posted up saying all Officers and warrant officers could
have leave ashore until 10 p.m. I went ashore with Capt. March at 3
p.m. and we walked around for a couple of hours and then hired an
"Aribeyah" for an hour and took a drive around the City. After making one or two small necessary purchases we went at 7 o'c to the
"Casino Palace Hotel" and had Dinner at 7:30 p.m. At the Hotel we
met two Royal Scots Officers who were on the *Alaunia* with, so the
four of us had Dinner together. It was after 9 o'c by the time Dinner
was finished, so we strolled up thro[ugh]' the City and went on
board getting thereabout 9:50 p.m. It is not much of a City tho far
ahead of Suez, but one can almost purchase anything in great variety.
However we enjoyed ourselves, it being a change from Suez.

Weather: Fine and warm and the evening was pleasantly cool, the
sea breeze being welcome.

P.S. — After arriving on board we had a regular high old time in
singing nearly everything we knew; some were what they call "jolly
with drink" and the jollifications were very amusing – however all
enjoyed themselves immensely for an hour and a half.

THURSDAY – MARCH 16/16
We left Port Said, for somewhere as yet unknown, about 8 a.m. Our
first Boat Drill Parade was held at 10 a.m. My boat was no. 15 containing 41 in all, and would be the first boat off those Davits. Our
Regt. was Duty Batn [Battalion] today; Capt. Alexander was Captain
of the Day and Lts. Rendall, Fox and myself was officers of the
watch. Our guard consisted of 80 men – 24 men being on Sentry, 1
on Signal Duty, 1 Bugler and 1 Orderly – at one time. All on board
had to do one hours Route march around the decks – about a couple
of hundred being on at one time.

Weather: Fine and warm with smooth seas.

FRIDAY – MARCH 17/16
Boat Drill and Route March as above. At 2 p.m. there was a demonstration to all officers of the best method of fastening on life belts and on the way the men were to get into the boats, which lasted an hour.

Being St. Patrick's Day there were general "carryings on" after Dinner. We had a general Court Martial; Lts. Nunns, Taylor, Summers and myself being Court Martialled. Afterward we had pillow-fights and general "shambles".

Weather: Fine and warm though slightly cooler.

SATURDAY – MARCH 18/16
Boat Drill and Route Marching as yesterday. Had another Court Martial tonight. This time it was Lt. Knight. Nothing Important today.

Weather: as yesterday.

SUNDAY – MARCH 19/16
Methodist and Presbyterian Church Parade at 9:15 a.m. and C. of E. at 11:15 a.m. – At 10:30 a.m. there was an Alarm Boat Drill Parade, there having been none at 10 o'c being Sunday.

At 3:15 we sighted the first land since leaving Port Said. It was Malta (about the middle of the morning we saw a little land on our starboard side, some say that it was a small island, others that it was Sicily) which we reached at about 4:30. We lay to about a mile off when we got our orders. We do not know where we are now going; whether to Marseilles, or to "Blitey" but it is likely to be the former. Tonight we had more fun – pillow-fights etc.

Weather: There was a very smooth sea today and the weather was perfect, being quite warm.

MONDAY – MARCH 20/16
This morning we were off Cape Bon on the African Coast and we could see land all morning. We had our usual Boat Drill Parade at 10 o'c and one hours Route March. During the afternoon we saw several steamers going in the opposite direction in which we were going. At night we had more pillow-fights and general "ragging".

Weather: Fine and warm.

TUESDAY – MARCH 21/16
Usual Boat Drill And Route Marching. We had a Kit Inspection also. At night we had another Court Martial, the accused being Lieut. Begg [NI] of the Royal Scots. He was very comical and amusing.

Most of us went to bed immediately it was over, being our last night
on board.

Weather: Fine but cool and windy.

WEDNESDAY – MARCH 22/16
We reached Marseilles about 7 a.m. Immediately after breakfast
arrangements for disembarking were made. I was given the job of
looking after the unloading of Ammunition, and when landed had to
place a Guard over it.

All the men and stores were ashore before noon. Just at 1 o'c we
were told that lunch would be served to officers at 2/6 each, and a
few minutes later we heard that as we were not leaving until 9:12
tonight, leave would be granted to officers until 6 o'c but that not
more than half of each Company's officers were to be away at one
time. So Capt. March and I at once went up the City and had lunch.
We then walked around a bit to see some of the sights. We bought a
few P.C.s and I sent one home and one to E[lla]. Just as we were
thinking of taking a street back we met Capt. Alexdr. [Alexander] so
we went back in a Taxi at 3:30 p.m. to let Knight, Bartlett and Reid,
the other officers of "D" Coy., off. At 6 o'c the Coys were "fallen
in", and we marched to the train – about 15 min – and the men were
fixed off in the Carriages. We are not leaving until 9:12 so have
about 2 ½ hours to wait. We believe we are going to Arras which is
on the other side of Paris, little more than half way from Paris to
Boulogne.

We left Marseilles at 9:30 p.m. and our first stop is to be about 3
o'c tomorrow morning.

Weather: Fine but chilly.

EXTRACT FROM POSTCARD DATED 22 MARCH 1916
Marseilles.
Passing through here. You can guess where we are now going.

THURSDAY – MARCH 23/16
The train stopped at 4 a.m. and we gave the men Tea, Bisquits,
Cheese and Jam. The Officers purchased lunch (of a kind at a Buffet
– I had a small cup of Coffee and a small slice of bread – at cost of 2
frs.[francs]. We left again at 5:30 a.m. and our next stop was at
Macon at 3:30 p.m. which we should have reached about noon. The
men were again given Tea etc and the Officers "fed" at a Buffet for 3
frs. We again started at 4:30 p.m. and stopped again to give the men
Tea, but there was nothing for the officers; we were 5 hrs late reaching this place.

We passed thro' some pretty scenery but the weather spoilt it to a certain extent.

Weather: Very cold and rainy.

FRIDAY – MARCH 24/16
Our next stop for food was at 4 a.m. and the next at noon when some of the officers got nothing to eat and others nothing – I was one of the latter, for, as the train stopped for only half an hour, by the time the men had their Tea the train was starting again. At 3:15 p.m. we had a shower of snow.

We stopped again at 7 o'c for half an hour to give the men a cup of Tea. – Nothing for officers. We reached the end of our railway journey at 2 a.m.

Weather: Rainy nearly all day and very, very cold.

SATURDAY – MARCH 25/16
Arrived Pont Remy at 2 a.m. and at 2:30 started to march to our Billets in Buigny [Boigny] l'abbé, which we reached about 3:30 a.m. having had a halt of ten min. when half way. On the way we had a small snow shower and it was very cold. An interpreter who had who had been detailed to billet us, now showed us to the various billets. "D" Coy's men were placed in Stables in lots of about 25 men. The Officers billet was a house of 3 rooms, the family who had the care of it living next door. As soon as the men were billeted they were soon asleep. We then went to see about the re-issuing of their blankets to the men (for they had been shipped together in bundles of about 25) and about 6:30 a.m. we had the men awakened and got them to fetch their blankets. The men again had a couple of hours sleep, and when things were straightened the officers got one hours sleep – all we had for the night.

About 10 a.m. we held a Rifle and Ammunition Inspection and inspected our men's billets to see that things were O.K. During the afternoon Platoon Cdrs. [Commanders] made their men acquainted with Rules and Regulations in regard to billets.

Weather: Cold and rainy.

SUNDAY – MARCH 26/16
I omitted to state yesterday that the C.O. sent for me and told me he wished for me to super-intend the Police, the prisoners and their work, in fact he wanted me to act or rather do the work of an A.P.M. [Acting Provost Marshal]. Consequently I had a busy day of it today as, after speaking to Provost Sergt. Greene, I found several things which should be managed differently, such as the feeding of the police, of the prisoners, etc. etc.

Service for Methodist and Presbyterian was held at 9:30 a.m. and for C. of E. at 11:15 a.m.

Weather: fine and cold, except for a couple of rain showers.

EXTRACT FROM LETTER DATED 26 MARCH 1916.
"Somewhere in France"
Am writing this several thousand miles from where my last letter was written, a fortnight ago.

The Censorship is, of course, much more severe here, and though I censor my own letters, yet I cannot say anything that would be of Military value, for fear the latter would fall into the enemies hands. When such matter is eliminated, it leaves practically nothing to say.

Though we have been here for only a couple of days, we are moving again in another of couple.

It is very cold but expect we shall soon get used to it again.

I hope all at home are well and may say that there is no need to bother about us at all, for we shall soon finish off this blooming war.

MONDAY – MARCH 27/16
We had a Battalion route March from our town Buigny l'abbé to Ailly-le-Hant [Haut]-Clocher, then to Youcourt [Yaucourt] Bussus and back home – a distance of 7 ½ miles. This was in preparation for a Brigade Route March tomorrow. In the afternoon we had an Inspection of feet each officer inspecting his own platoon, also a Rifle inspection.

Weather: Fine and cold.

TUESDAY – MARCH 28/ 16
We had to get up extra early this a.m. for we had breakfast at 6:30 a.m. and parading at 7:30 a.m. as all units had to assemble at Brigade Hd. Qtrs. at 9 a.m. which was 3 ½ miles away. After leaving our town we marched through Bauchelles-les-Quesnay, St. Requier, Yaucourt-Bussus and back home – a distance of 10 3/4 miles, arriving back at noon.

In the afternoon the officers again inspected their men's feet, also Rifles, Ammunition, etc.

Weather: Fine and cold.

WEDNESDAY – MARCH 29/16
Company Parades today. D. Coy Platoon officers lectured their men on the uses of the Gas Helmet with hints on effectiveness.

In the afternoon all mens blankets and officers Kits were packed and loaded on transports for practice. Officers were allowed to retain only 40 lbs including blankets, the remainder was sent to the Stores for shipment to England.

Weather: Rainy and cold.

THURSDAY – MARCH 30/16

Company Parades. Gas helmet Lectures by Platoon officers and Inspections and Platoon marches etc. In the afternoon we had a short Company March. About 7 p.m. orders were issued that we were to march to new billets tomorrow.

Weather: A perfect Summer day in all respects.

FRIDAY – MARCH 31/16

Inspection of Platoons at 9 a.m. also of Billets, etc. At 11 a.m. we started for our new billets at – Brucamps – 6 ½ miles. We passed through Ailly-le-haut Cloucher and the Batn. reached Brucamps about 2 p.m. When a mile and a half from our destination just after dinner for the dinner had been cooked en route I was sent back to pick up some men who had dropped out. I found a couple before reaching Ailly-le-haut Cloucher, but knew there were a couple more who had dropped out before the battalion had reached A. le h. C. and as I could not find them decided to go back to Buigny L'abbe to see if they had gone back. As it was 3 ½ miles back I got the loan of a horse from the Royal Scots who were then billeted at A. le h. C. and rode back. I tried to converse with some of the inhabitants and eventually found out that two men had been around there for the past two hours. After looking around for awhile I saw them in the distance. They also saw me and made off. I gave chase at a gallop. They made into a farm yard and thro a gateway with a bar across about a foot from the ground, so I gave the horse straight for it and he got over it O.K. and I eventually got near enough to recognize one of them so called to him by name to stop. They were then going thro' some hedges, but he stopped and called to the other to stop. I then made [them] walk back with me, and when reaching A. le h. C. returned the horse to the Royal Scots, and continued the journey on foot picking up at A. le h. C. a man who had come with me as an escort. Having heard that one of our men had been seen going down a road which was not the right one, I sent the prisoners I had along the right road and went to look for this other man, but could not find him though I went 2 miles along the road. I then came back again and caught up to the others just as they were entering Brucamps, about 6:30 p.m. and reported to the C.O. On the way to Brucamps we picked up the Equipment and Rifle of a man who had fallen out and followed the way the Batn.

had gone, later – he was slightly drunk, and being a Lce. Cpl., lost
his stripe next day.
 Weather: Fine and warm.

SATURDAY – APRIL 1/16
Left Brucamps at 9:30 a.m. for Bonneville. Before starting the C.O.
asked me to march in the rear of the Batn. and take note of all men
who fell out and make any who were not bad enough to fall out
return to the ranks.

SUNDAY – APRIL 2/16
I was detailed for the C. of E.[Church of England] Service at 11 a.m.
In the afternoon we had Platoon inspection.
 Weather: Fine and warm though cloudy.

EXTRACT FROM LETTER DATED 2 APRIL 1916.
The Machine Gun sections of the various battalions in our Brigade
have been formed into a separate Company, the Machine Gun Co.;
and as there was one section too many in our Brigade, our section
was returned, which meant the abolishing of our Machine Gun Company. The reason ours was returned was principally on account of the
difference in pay of our men and that of regulars.
 To-morrow morning at 6.30 a.m. Capt. March, O.C. 'D' Company
is going off on leave for the period of his absence I shall not be able
to go until he comes back. Am trying to arrange for Jim and I to go
together. Some of the Sergeants and most of the officers took holidays (five days) to Cairo when we were at Suez, but Jim and I did
not.

MONDAY – APRIL 3/16
Company Parades today. We had Gas and Helmet Drill principally,
with a little musketry, Etc.
 Weather: Fine and warm.

CHAPTER SEVEN

Louvencourt to Eve of Battle
(4 April 1916 – 30 June 1916)

The British Army that collected on the Western Front in the spring of 1916 was the largest force ever assembled in that country's long military history – no British general had ever commanded so large a force in the field. Comprised of volunteer units raised in response to Lord Kitchener's injunction that "Your Country Needs You," this army had no field experience and represented little more than a collection of divisions. The army commander, Sir Douglas Haig, was understandably hesitant about committing this untrained and inexperienced force against a battle-hardened enemy. Nevertheless, circumstances convinced him that it was necessary. French manpower was already overextended and Haig felt that before long the brunt of the war in the west must fall on British Empire forces. As well, Allied plans for a coordinated offensive on all fronts held the promise of an early victory.[1] On the Western Front a massive Anglo-French attack along a sixty-mile front on the Somme in late June or early July had been expected to roll back German lines and lead to victory. But Germany's unanticipated and ferocious attack on Verdun (February-December 1916) reduced France's initial participation in the July drive and forced major changes in the original plan. Britain's inexperienced army would now bear the weight of the offensive.

General Sir Edmund Allenby's[2] Third and General Sir Henry Rawlinson's Fourth Army reinforced the British line. The Fourth Army's fourteen-mile front, which included Major-General Sir Beauvoir de Lisle's 29th Division, extended from Maricourt, where it formed a junction with the French Sixth Army, to a half-mile south of Gomme-court, a village just inside the Third Army's front. The enemy held the whole of this front strongly, with a three-tiered system: forward trenches, well dug in with extensive protective wire and capable of surviving sustained bombardment by hostile artillery, beyond that, at ranges of between 2000 and 5000 yards, a second line of trenches and, farther back, yet a third was being constructed. The German network of heavily defended lines presented a formidable obstacle to any attacking force.[3]

The 29th Division, comprised of the 86th, 87th, and 88th Brigades, occupied the forward slope of the most westerly of a series of well-defined ridges that projected south-eastwards toward the Ancre River. In a shallow intervening valley lay the little village of Beaumont Hamel, just within the German front-trench system and overlooked by a minor projection of the Auchon-Villers ridge, Hawthorn Ridge, that reached towards No Man's Land. The Newfoundland Regiment, as part of the 88th Brigade, moved into the firing line on Easter Eve, 22 April, the first time they had been in the firing line since Gallipoli. Some 300 to 500 yards distant, down a grassy slope lay the German front lines, their heavy barbed wire entanglements guarding the approaches. The German 119th Reserve Regiment, tough and experienced, had turned the natural defences of a deep Y-shaped ravine into a veritable fortress, one of the strongest positions on the entire Somme front. For the next two months the Newfoundlanders shifted between the forward area and the reserve.

The move to the firing line entailed lengthy and arduous preparation of forward positions dangerously exposed to enemy fire. Over the course of the next two months the terrain and the names connected with it would become familiar to Steele and the other Newfoundlanders. It was risky work and resulted in the first casualties since Gallipoli. The German lines were only a few hundred yards away in places, and the bombardment of Allied positions was frequent and nerve-wracking. For Steele, it raised some concerns about being wounded, especially before he went on leave to England with Jim. He returned on 12 May to take up his duties as Billeting Officer and Acting Company Commander. Kept busy with preparations for "our Great Advance," Steele had little time for writing and his diary entries shorten considerably.

Lord Kitchener's untimely death on board HMS *Hampshire*, sunk by a German submarine,[4] unsettled Steele and his compatriots. So, too, did the news that the Battle of Jutland had not secured a British victory. On the surface Steele exuded confidence that the upcoming battle would end the war and bring on a lasting peace. The detailed preparation for the offensive had convinced many that the Division was well prepared for battle. But as the day of battle approached Steele seemed less confident. Perhaps it was because he knew that the enemy was expecting an offensive and was busy strengthening its defensive positions: the "Big Day" would hold few surprises for an alert enemy.

The original plan was that the offensive would begin on 24 June a day of special significance to Newfoundlanders for on that day in 1497 John Cabot laid claim to the New Founde Land on behalf of the English King Henry VII. Had it taken place as planned the irony would have been remarkable – the country's regiment devastated on its national day. But conditions were not right and the offensive was postponed for another week.

Just a week before the battle Steele was injured when his horse shied at a dead horse and drove his leg against the wheel of an approaching wagon. The doctor wanted him to be hospitalized and informed Lt. Col. Hadow accordingly. Steele demurred, fearing that he would be the subject of ridicule and accused of "slinging the lead."[5]

This final week was a busy one as detailed preparations for the offensive began in earnest. The 29th Divisional order had been specific in the equipment each soldier was to carry with him into battle: rifle and 170 rounds of small arms ammunition, one iron ration and rations for the day of assault, two sandbags, two Mills grenades, steel helmet, smoke helmet in satchel, water bottle and haversack on back, first field dressing, and identity disc. The identity disc, a triangular piece of metal, seven inches a side, cut from a biscuit tin, was to be fastened on each soldier's back, to enable air reconnaissance to spot the advancing columns on the day of battle. Unfortunately, it also provided a ready target for German sharpshooters. Those in the first and second waves carried an additional 120 rounds of ammunition. At least 40 per cent of the infantry carried shovels and 10 per cent picks. In addition, the regiment had to carry its share of equipment issued by the brigade, including flares, five-foot wooden pickets, mauls and sledge-hammers, wire-cutters, hedging gloves, haversacks and bangalore torpedoes.[6] A further major item were the wooden bridges for crossing trenches.[7] On average this meant that the men would carry an extra sixty pounds of equipment into battle, nearly half the weight of the ordinary soldier at the time.

The Newfoundlanders were confident that they were well prepared. Weeks of hard training carrying the bulky equipment needed on the final assault had conditioned them physically. Endless practice over the same terrain had made them comfortable with the task assigned. General de Lisle spoke to the regiment on the 26th, offering convincing evidence of the power of the British offensive and the weakness of the German position.[8] This was to be "the greatest battle in the history of the world," recalled Jim Steele four years later.[9] Few doubted the outcome. "Tell all friends that the 1st Newfoundland are O.K., and never feel downhearted. We will make you all proud of us some day," wrote Private Mayo Lind on the 29th. It would be his last letter home.[10] Realistically, there should have been a greater sense of foreboding. German air reconnaissance over the 88th's sector had intensified throughout June and enemy artillery fire was busy with its registration[11] of divisional front lines. Both indicated that the enemy was aware of an impending attack. Two nights in a row Newfoundland raiding parties crossed out into No Man's Land heading for the German trenches seeking intelligence on the enemy and prisoners to bring back for questioning. The first night they were unsuccessful. They were more successful on the second, learning something of the enemy's strong position and high state of readiness. No prisoners were taken.[12] Reconnaissance patrols by other units in 8th Corps, of which the 29th Division was a part, confirmed the Newfoundland reports. And yet General Headquarters (GHQ) ignored this important intelligence. Sir Douglas Haig focused instead on the failure of the raids to break into the German positions, a failure which he attributed to poor staff work.[13]

The regiment was in great spirits when they marched to the front lines on the thirtieth.[14] There were a total of 25 officers and 776 NCOs and men present, including 66 green reinforcements who had arrived from base that afternoon. Left behind under the command of Major Forbes-Robertson,[15] on instructions from GHQ, were the "ten per cent" intended to form the nucleus of a revitalized unit in the event of severe casualties. A certain percentage of officers, including the battalion and company second-in-command, was normally held back. Steele, as second-in-command of D Company, was part of this reserve, perhaps because of the injury he had sustained earlier.[16] This decision appears to have been made fairly late in the preparations as Steele makes no mention of it in his diary or letters.

Entries for the last week before the battle contain only two letters, both of which reveal some of the foreboding Steele felt as the preparations intensified. Time seemed to be passing too quickly. Unexpected delays in commencing the "Great Push" and troublesome news about the enemy's heightened state of preparation intruded upon Steele's otherwise confident mood. The entries reveal a strengthening of emotional ties with home – thoughts of being able to spend Christmas with family should the

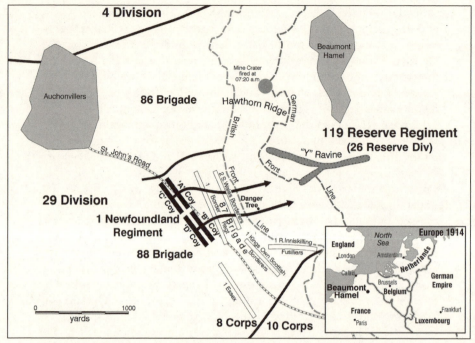

Beaumont Hamel, 1 July 1916

offensive be successful. His final diary entry is the most poignant: the recurrence of the word "last" says much about his apprehension as the regiment prepares to move off. The last page of his diary, with the heading, Saturday, July 1/16 is blank. A single page with the numbers of casualties for his company had been inserted but there are no comments.

TUESDAY – APRIL 4/16
Changing Billets again today, moving Louvencourt.
Left Bonneville at 9:30 a.m. passed thro' Caudan [Candas], Doullens, Freschevillers, – touched Ordille [Orville], passed thro' Sarton, Marieux, Bauchelles-les-Authie [Vauchelles-les-Authie] and reached Louvencourt, having walked about 16 miles. We had our Dinner on the way it being cooked in our field kitchens en route and was really excellent.
Weather: Fine but cool, being perfect marching weather.

WEDNESDAY – APRIL 5/16
Company Parades – we had general inspection, arms, equipment, etc. At 2 p.m. all B Co. and 55 men of D Coy. with all officers of both

Companies were warned to be ready to march to Lealvillers at 5:30 p.m. to work at Railway Construction. After one or two changes in orders 55 men were to go from "D." Co. Half an hour before starting I received orders not to go as I was required for a Court Martial tomorrow. They left at 5:30 Capt Donnelly in charge with Lts. Knight, Bartlett, Reid, Shortall, and Nunns. Capt. March did not go because he was going on his leave tomorrow morning.

I had to act as Prosecutor at the C.M. [Court Martial] of Sergt. [illegible] tomorrow at Bde. Hd. Qtrs.

Weather: Fine – little cool towards evening.

THURSDAY – APRIL 6/16
Capt. M. and Alex. – Lts. Butler and Stick and the c.o. with 15 men left this a.m. at 6:30. I was on the c.M. all morning. In the evening I got orders to go to the firing line tomorrow morning with Lieuts Paterson [Patterson], Herder[17] and Strong to have a look around.

Weather: Cloudy, and cool in the evening, though fine.

FRIDAY – APRIL 7/16
Had to go to the Court Martial this a.m. for about an hour. About 11:30 a.m. Herder, Paterson [Patterson], Strong and myself started for the firing line. We each had a horse – I had the Doctor's [Lt. Frew's] – so had quite a pleasant run up, for it was about 7 ½ miles of a ride and 3 miles of a walk. There was nothing at all alarming or wonderful about it – in fact, I consider our Suvla Bay line as good as if not better than the part we visited. We got back to our billets at 6 p.m. having walked about 7 miles and ridden about 15 miles.

After I had made yesterday's entry I received word that the Adjutant wanted me. He wanted me to go to Lealvillers, nearly 3 miles away, where B. Co and most of D. Co were working at Railway construction work under Capt. Donnelly and the other 5 officers. Our party was supposed to relieve the R.F.R. [2nd Bn Royal Fusiliers] but had not done so, and several letters were received by the Capt. from the Brigade and the Divn. on the matter. Then the division said we were to send an officer immediately to find out what had been done & to have the matter straightened out.

I left at 8:30 p.m. and it was midnight when I got back, having arranged for our party to relieve at 8a.m. Just when I started home heavy bombarding started in the firing line (it was in the same part which we visited today) and gradually increased in violence until there was one continual roar cannonading and the sky was continually lit up. Bursting shrapnel could continually be seen. It did not slacken in violence until 11 p.m. and in another half hour except for an occasional report it had assumed normal. We have been in parts of these trenches today and saw a few dead men whom they had not yet buried, and several sections of trench which had been broken up by Artillery fire. It seems that one of our patrols met a German patrol and they started bombing, etc. each other, which developed to the above extent. Some of the Germans came over to our trench and took away some prisoners, some reports say only two or three, others, a whole platoon; the Germans were eventually driven back.

Weather: Fine during day, but a little cool and a little showery at night.

SATURDAY – APRIL 8/16

We had a route march of about 8 ½ or 9 miles. We left Louvencourt and marched thro' Bauchelles, Marieux, Sarton, Thievres, Authie [Vauchelles-les-Authie] and back to Louvencourt. There was nothing doing during the afternoon.

Last night a Draft of 120 men with Capt. Ledingham and Lt. Pippy arrived. They had been to Alexandria in Egypt by error and had come here via Marseilles. Four other officers went to Marseilles – Keegan, C. Rendell, Robertson, and W. Ayre,[18] but had been kept by the Base in Alexandria as they were short of Officers.

Weather: Fine and warm.

SUNDAY – APRIL 9/16

Church Parades this a.m. In the afternoon I went down to Lealvillers where our working party was, to see about some Company matters and to see Jim, then came back to Louvencourt and went to Acheux for a run.

Weather: fine and warm.

MONDAY – APRIL 10/16

Company Parades to-day. Capt. Ledingham and Lt. Pippy were attached to details of B and D Coys, the former being in charge.

At 5 p.m. after short notice we paraded in full marching order and marched to Englebelmer and continued to and occupied the third line and remained there for an hour and a half, and whilst there were visited by the Brigadier and Brigade Major. At 10 p.m. we reformed and marched back to Englebelmer same way as we came – thro' Force-vielle and Acheux (This was for practise).

Weather: fine and warm – little cool at night.

TUESDAY – APRIL 11/16

Company Parades today. Helmet – Musketry, etc., etc.

At 3 p.m. the battalion – or what of it was here, attended a lecture at Brigade Hdqtrs. on the use and care of the Gas Helmet for gas attacks. We then marched to a shack where two cylinders were keeping the place filled with gas and putting our helmets on everyone passed thro the place. I remained there for ten mins. but felt no effects whatever.

Weather: Fine in morning but rainy in afternoon.

WEDNESDAY – APRIL 12/16
Company Parades.

As it was raining we kept the men in their billets and lectured them on Gas Helmets, March, Trench, and General Discipline etc. and also inspected all helmets and gave the men a second helmet.

Weather: Rainy all day.

THURSDAY – APRIL 13/16
Inspection Parades during morning and getting ready to move to Englebelmer [3 kms from the front line]. About 3 p.m. the detachment working on Railway Construction at Lealvillers and Acheux returned. Cpt. Ledingham and Lt. Pippy were both attached temporarily to B. Co. At 5:30 p.m. we started – (I was in command of "D" Coy. owing to Cpt. March being on leave). We marched thro Bertrancourt, Beausart [Beaussart] and Mailly Maillet and reached Englebelmer about [omitted] having marched about [omitted] miles. By the time we had the mens blankets issued to them and all the Coy. billeted it was 11 p.m.

Weather: Fine but cool.

FRIDAY – APRIL 14/16
Today was spent in getting the men straightened off in their new billets and in having one or two small inspections, such as of Rifles, Ammun[ition]., Helmets, etc.

Weather: Rainy most of the day.

SATURDAY – APRIL 15/16
We had various equipment inspections today and had quite a bit of work getting ready for tonight's working parties. There were three reliefs, 7:15 p.m., 9:45 p.m. and 12:15 a.m. A Co. had the first relief, B Co. the second and D Co. the third whilst "C" Co was divided up between A, B, and D. Coy. Capt. Ledingham was in charge of A. Co. (he being now attached to A. Co) Capt. Donnelly of the second and I of the third. We were making a Communication Trench – it has been called "Tipperary Avenue" – all communication trenches are named.[19]

We arrived back to our billets at 4:30 a.m. and by the time we got the shovels and picks put away it was 5 a.m.

Weather: Raining at very frequent intervals.

P.S. A new draft of about 200 men with Lts. Molloy [NI] and Cashin[20] arrived tonight. The C.O. returned also.

SUNDAY – APRIL 16/16

Sunday Services today and the customary inspections. At night we were on the same work as last night.

A day or two ago all officers and men (of the 29th Divn at least) were cabled to return to their units at once, so it seems as if something is brewing.

Weather: Quite a fine day but rainy at night.

P.S. The C.O. inspected the men of the draft at noon, they had been divided among the 4 Coys.

MONDAY – APRIL 17/16

We had our small inspections again today and at night continued work on "Tipperary Avenue".

Capt. March and three or four others of our officers and a dozen or more men returned. Some of them including Capt. March had had all their leave.

Weather: Raining at frequent intervals in the afternoon and night though the morning was fine.

TUESDAY – APRIL 18/16

Company Inspection Parades as usual. Capt. M. takes over "D" Co. Today we received orders that we were to put five hundred yards frontage of barbed wire entanglement in front of the support line near Auchonvillers. The method of doing so was demonstrated to the officers in the afternoon. We then showed our men. We left at 7 p.m. and reached home about 3 a.m. having accomplished our work.

Weather: Rainy nearly all day and practically all night. Very cold and very disagreeable.

WEDNESDAY – APRIL 19/16

Company inspection parades. At night we worked at "Tipperary Avenue" again and finished the part allotted to our battalion for the night. We had only two reliefs tonight, for we had only two Companies, A and B Coys having gone into the support lines last night. They were supporting the Worcesters who were in the firing line in front of them. Today we had one casualty (Cooper "B" Co.),[21] the first since we have come to France. He had a "cushy"²² one in the shoulder.

Weather: Rainy most of the day and night.

THURSDAY – APRIL 20/16

Usual Company Parades.

Today I received word from the c.o. to take over today the duties of Billeting Officer as Fox was going to Hospital tomorrow – his heart is again giving him trouble. The c.o. told me that he wanted me to attend to Company work also when I was not engaged in the Billeting. I told him I also had the Prison department to attend to; he had forgotten this but told me to do the best I could as we were short of Officers. About 8:30 p.m. orders were given for C and D Coys. to go to "Tipperary Avenue" at 2 o'c tomorrow morning.

Weather: Rainy nearly all day but clearing at night

FRIDAY – APRIL 21/16

At 2 a.m. we all got up and all went to "Tipperary Avenue" for 4 hours work at draining a certain portion of this Communication Trench. It was a glorious morning though very cold. Whilst engaged at this work the sky was continually occupied by Aeroplanes, about five or six being up at one time. We were unable to decide whether they were our own or German, though they were shelled almost continually. There must have been about 5000 anti-aircraft Shells sent up at them. I counted 150 fired at one of them in less than one minute.

We reached back to our billets at 8:30 a.m. The Adjutant told us that we were leaving these billets tomorrow evening as C and D Coys were going into the Firing Line. I therefore had to spend the day in getting my billeting accounts made up. The two Coys. went in again at 1:30 p.m. and returned at 6:30 p.m., and am glad I was unable to go, for it [rained] practically all afternoon. The mud in the trenches has been terrible lately. I have my long rubbers with me which I find very useful just now, and this morning when in "Tipperary Ave." I had the pleasure of getting into the half mud and water over the top of them by about 3 inches, and I had to take them off to get the liquid mud out of them.

Weather: Fine in morning but raining afternoon and night.

P.S. It is 12 mos. today since I received my Commission.

SATURDAY – APRIL 22/16

The men did no work today, as we were to go into the Firing Line tonight. At 8 a.m. all officers and Sergeants went to look over the portion of the Firing Line which we were to take over. We, "D" Co.

had about 400 yds. A and B Coys. were remaining in the Support
and were supporting "D" and "C" respectively. When we returned,
about 2 p.m. we were mud nearly all over and were mud-soaked to
about three inches above the knees. I had a pair of long rubbers but
they were filled with liquid mud.

After having lunch I had to go to Acheux on my horse[23] with the
Interpreter to fix up the billeting of the transport to date. I got back
about 6:30 p.m. having ridden about 16 miles – it was raining all the
time. At 7 p.m. the two Coys. C. and D. went in to the firing line and
I and my orderly went in about half an hour later, and got in as
quick as the Coy. – it was an hour's walk.

Raining all day but clearing towards midnight.

SUNDAY – APRIL 23/16
Easter Sunday. Today opened a perfect day – a real Easter Day to be
sure.

The men spent the day repairing the firing line which was certainly
in very poor condition, and though held by the British since last
August, were not to be compared to our Suvla Trenches[24] in any way.
They were small, firing steps and traverses and parapets all poorly
made.

After breakfast I went back to Englebelmer to see to the billeting of
those who were remaining behind and also of some of the transport
horses and men who were coming up from Acheux. Arrived back to
our lines about 6 o'c. Things were very quiet tonight except for an
occasional burst of shelling by the Germans which had no result.

Weather: Perfectly Summer like.

EXTRACT FROM LETTER DATED 23 APRIL 1916
France.
So this is Easter Sunday and a beautiful day it is, though the past
week has been nothing but rain, rain, rain. Consequently, there is
mud everywhere. It is about the consistency of *very thick* 'pea soup,'
and is of a reddish brown colour in most places. It recalls memories
of Suvla.

We went into the Firing Line last evening and I have come out this
morning to attend to some duties in connection with a new appointment I have received, so am taking the opportunity to drop you a
couple of lines.

Two or three days ago Lieut. Fox went to hospital, being again
troubled with his heart. He was 'Billeting Officer' and on the day he
went to hospital the Colonel appointed me in his stead. When the
Battalion is going to change its billets, the B.O. [billeting officer] goes

to a new village a day ahead to arrange with the "Town Comman-dant" for the billeting of the officers and men. Then the B.O. has to square up all accounts in each village (by means of vouchers) and has to make up just how much is owed for men, officers, horses &c. for each of whom there are different charges. When the billeting officer goes to prepare new billets, he takes with him the Battalion Inter-preter, (there is one attached to each Battalion), four cyclists to guide each of the four companies to their places and one man on horse back for the transport. Both the Interpreter and myself have a horse. Yesterday we rode nearly sixteen miles. Besides having this work, I have, when free of billeting work, to attend to my Platoon, so I told the C.O. I also had the prison department work.[25] He had forgotten this so he told me to do the best I could as we were short of officers. You will thus see that I shall now be busier than ever, which is saying something; However, it is part of my "little bit."

Jim and I are both well and getting along all right and more or less happy and contented.

MONDAY – APRIL 24/16
Everything still very quiet, men engaged on improving our Firing Line. At night work was done which involved the throwing of earth on the parapets as this could not be done in the daylight. One man was killed in C. Coy. today – he came in the last draft. It was his own fault, for he had been half up on the parapet today and was continually warned to get down. His name was Curnew[26] and he had a brother in D. Coy. who was through Suvla. Three men were wounded through the day including Lt. Cashin of "B" Co.

Weather: Fine, warm, and Summerlike.

TUESDAY – APRIL 25/16
Everything quiet again. There seems to be very little doing in our part of the line just at present. Generals De Lisle and Caley and Staffs were through our lines today and said they were well pleased (for a wonder) with the condition of our trenches and the work we had done. At eight the Germans gave us a couple of bursts of shrapnel fire and though no casualties occurred there were several narrow escapes. One man in my platoon had his life saved by his steel hel-met; a piece of shrapnel passed through the rim of his helmet and landed on back of his neck, but he hardly felt it. Another man in 16 platoon had his foot grazed and returned to duty after visiting the Doctor.

Weather: Fine, warm and Summerlike.

WEDNESDAY – APRIL 26/16
Major Drew returned to us today – no one expected him to return.
By the way 2nd Lt. W. Ross,[27] who only came on Sunday Evg., went
to Hospital, sick, yesterday. Usual work – Very little doing in the
Shelling line.
    Weather: Fine, warm and sunny.

THURSDAY – APRIL 27/16
Today I received six letters which had been to the Mediterranean and
back.
    The men were on the usual trench work all day.
    Weather: Fine, warm and sunny.

FRIDAY – APRIL 28/16
Today is my [twenty-ninth] birthday and this time last year I was in
Edinburgh Castle.
    Our side was doing the usual amount of intermittent shelling but
the Germans were unusually quiet.
    The men were still on trench improvement work.
    Weather: Fine and warm.

EXTRACT FROM LETTER DATED 28 APRIL 1916
In the Firing Line, France.
We have been in the Firing Line for the past few days. I regret I can
give no particulars of our stay there, but we had a very quiet time.
    Regarding Socks:- Jim and I have sufficient for some time thank-
you.
    All the boys are keeping well here in France, very few going to
Hospital. Lieut. C. Clift went to Hospital about a week ago, and a
couple of days ago Lt. W. Ross went, after spending only one night in
the Firing Line, having only just arrived here from Scotland; He is a
brother of Lieut. H. Ross who was wounded at Suvla last November.
    Our Firing Line where we are at present is a very poor one. The
German Lines are about two to five hundred yards away, with nothing but grassy fields and lots of barbed wire entanglements. Rifles are
very quiet, but shelling is frequent and plentiful and very heavy. We
have very narrow escapes sometimes from "Jack Johnsons," "Coal
Boxes" and other large High Explosives.
    Jim and I have not yet had our leave but expect to get it any
day.
    There are any amount of Aeroplanes here, but we predominate,
There are nearly always from one to a dozen up and the opposite
side always sends hundreds of anti-aircraft shells up after them.

SATURDAY – APRIL 29/16
Trench improvement by day and at night working on a new
advanced firing line.
Weather: Fine and warm.

SUNDAY – APRIL 30/16
Same work as yesterday. At night I went out with the working party
for the advanced firing line. I first of all took out Corpl. Bellows[28]
and five men as a listening post, their spot being near some trees
nearly 300 yds. in front of our present firing line, and the same from
enemy so that they could watch and listen if there was anything
doing. I then placed a "Covering Party" for the working party; the C.
Party were about 25 yds. in front of the new firing line where the
working party was. When this was done I placed the working party
and started them on their work.
Weather: Fine and warm.

MONDAY – MAY 1/16
Trench improvement work. At night I placed the same parties as last
night and just as I had the working party ready to start the Germans
commenced to bombard our Brigade's Lines, particularly the Hants
and our own. It got so bad that we had to return to main firing line
so as to get more shelter. Just as I reached back I got a note that had
been sent to me some little time before by the Adjt., that I could go
on leave and that I could go to our stores in Englebelmer to prepare
immediately. As one other rank could go I got Capt. M.[arch] to let
Jim go. Of course we could not go until the bombardment was over,
and I felt every minute he or I might be wounded, and thus interfere
with our holiday. Whilst it lasted it was very severe and nerve rack-
ing, indeed, it was the worst we had been in. Tho' it lasted for about
an hour and a half we were once again justified worthy of the desig-
nations "Lucky", for though the Hants had nearly a dozen casualties
we had not a single one still many times we were covered with debris
from shell bursts.
     After it was all over Capt. M.[arch] desired me to put out the cov-
ering and working parties before leaving, so it was 11 o'c when we
left the trenches for "Blitey". We went to the stores and left what we
did not want and took what we wanted.
Weather: Fine and warm.

TUESDAY – MAY 2/16
At 2:30 a.m. Jim and I met the Adjt. and the other Owen [NI] and
we started for Acheux, where we arrived at 3:45 a.m. and left by

train at 5:00. Havre was reached at 9 p.m. and we left for
Southampton at midnight on board the Irish Channel Str [Steamer]
Connaught.

The run through from our Sector was indeed enjoyable, for everything was green with thousands of apple trees in bloom everywhere.
The prettiest place I saw was Rouen – trees in various colours dotted
all through the City.

Weather: Fine warm and Summerlike.

WEDNESDAY – MAY 3/16
We were unmolested by Submarines during our trip across.
Southampton was reached about 9 a.m. and about 9:30 we entrained
for London, arriving about 11 a.m. After a visit to the Pay Office Jim
and I left for Halifax at 4 p.m. arriving Greetland Stn. at 9 o'c. and
Bowers Hall Barkisland at 10 p.m.[29] They were expecting us, being
outside, for I had sent a telegram from London.

Weather: Fine and warm.

MAY 4 TO MAY 9/16
We spent this period enjoyably at B.H. [Bower Hall] – Weather: Dull
and rainy practically every day.

WEDNESDAY – MAY 10/16
Left Halifax at 9:30 a.m. and arrived London 2 p.m. Left for
Gravesend at 4 p.m. and arrived via Tilbury Docks at 5:30 p.m.
They were of course glad to see us – I had sent them a wire
yesterday.

Weather: Fine and warm all day.

THURSDAY – MAY 11/16
Left for London at 10:30 a.m. and arrived 11:30.

Left London for Southampton at 4:30 p.m. arriving there at 6 p.m.
and went aboard Irish Channel Str. "Viper" – 27 knotter and left at
7:30 p.m.

Weather: Fine and warm.

FRIDAY – MAY 12/16
Reached Havre safely at 3 a.m. and left by train 4 a.m. – arrived
nearest Station to destination 8 p.m. and walked to nearest village –
one mile – where we found out that the Batn. was at Louvencourt.
Raley and I were offered bicycles so rode the 3 remaining miles.
Arrived to our billets about 9:30 p.m.

Weather: Dull but warm.

EXTRACT FROM LETTER DATED 12 MAY 1916
France.
Am writing this on the train at a station not quite half way to our destination, where it has stopped for an hour for us to get something to eat. The officers go to a Hotel a short distance from the station platform, and the men get supplied on the station platform. I have had my 'feed' and as I have half an hour to spare, am commencing this letter.

It is now 11 a.m. and we expect to reach our destination about 1 or 2 o'clock.

We have had practically nine day's leave, starting from the time of arrival in London to time of our leaving again.

Train is going again now, so cannot write more for the present.

SATURDAY – MAY 13/16
Rose at 5:30 a.m. and left for Acheux at 7 o'c with Lt. Ayre and a party of 100 men for various fatigues – 3 miles – returned at 4:30 p.m.

Weather: Rain – Rain – Rain.

EXTRACT FROM LETTER DATED 13 MAY 1916
We arrived back last night.

To-morrow morning I am taking over Billeting Duties from Captain Ledingham, who attended to them during my leave, and have to arrange for the re-billeting of the Transport and Transport men, and have also to arrange for further accommodation for the Police and prisoners. Then I have the prisoners's book to bring up to date. This means I shall be busy most of to-morrow. The Adjutant has just sent me a note to take charge of the Church of England Parade, which I consider too much of a good thing.

Sergeant Goodyear[30] has received a Commission and is now Transport Officer. He was Transport Sergeant.

SUNDAY – MAY 14/16
Being Sunday we had no fatigues, so I spent the morning taking over the Billeting duties from Capt. Ledingham, who had attended to same during my absence. In the afternoon there was a special parade when Genl. Caley of the 88th Bde. presented the medals to the Caribou Hill veterans: Sergt. Green[e] and Pte. (now Lce. Cpl.) Hynes.[31] Capt. D[onnelly] was getting his in England.

Weather: Fine and warm.

MONDAY – MAY 15/16
Same work as Saturday May 13th.

Weather: Raining all the time.

TUESDAY – MAY 16/16
Same work as yesterday.
    Weather: Fine and warm.

WEDNESDAY – MAY 17/16
Same work as yesterday. On returning to Billets, I got word from the
Adjutant to go to Englebelmer and Acheux and prepare billets for
Quartermaster Stores and Regimental details at the former and the
Transport at the latter.
    I went to Acheux on my horse at once but left Englebelmer for
tomorrow.
    Weather: Fine and warm.

THURSDAY – MAY 18/16
Went to Englebelmer at 9:30 a.m. a distance of about 8 miles, on
horseback and arranged the necessary billeting. Got back to Louvencourt at 4 p.m. and left about 5:30 p.m. for Englebelmer again to
join my Coy. – the batn. having gone early in the afternoon.
    Joined to E Coy. when they were entering the Trenches, at 9:30
p.m. It was Midnight when we got the Coy. put away. Capt. M.[arch]
and I could not find any dug-out, so slept – or tried to – on the firing
step – we were in supports. We could not sleep for we had no blankets, our orderlies not turning up.
    Weather: Fine and warm – cold at night.
    The C.O. went to England tonight to receive his Decoration.[32]

FRIDAY – MAY 19/16
We succeeded in getting a small dug out today, tho' not the one
which is ours by right. The Hants are using ours, so Capt. Ayre spoke
to the Major (acting C.O.) about it and he was seeing what he could
do about it. We had a few fatigues today, but not a very busy time of
it. Capt. E. Ayre arrived today and was attached to D. Co.
    Night arrived and we had heard nothing about our dug-out so
slept in same place as last night.
    Weather: Fairly fine with a shower of rain.
    P.S. The major (acting C.O.) told me today that as Capt. M.[arch]
was going on 2 mos. leave to Nfld. on private business, he wanted
me to take the Coy. until Col. Hadow returned, and that Capt.
A.[yre] would work under me to pick things up in connection with
the Company.

SATURDAY – MAY 20/16
Today the Coy. was on various small fatigues. We did not have a very

busy day – I spent the day getting into the details of the Company,
having taken it over last night. Capt. M.[arch] was going tonight, but
the c.o. asked him to remain until tomorrow.

Weather: Dull, but later, fine and warm.

SUNDAY – MAY 21/16
We had fatigues, – trench-digging etc., – all day and night, the men
having to put in four hours by day and four hours by night. This
morning I went to the Major and told him we had not yet got our
Coy. Hdqtrs., and that I needed it very much for my work. I had
already visited the Regt. occupying it (Hants) and asked him to back
it up by phoning, which he did.

However, we moved into it at 4 o'clock p.m. We will now be fairly
comfortable for we now have beds (made with sticks and wirework)
a table and two or three chairs. Capt. M.[arch] left about 3 p.m.

Weather: Very fine and warm.

MONDAY – MAY 22/16
Working on Communication Trenches etc., evidently preparing for
our Great Advance.

Weather: Fine and warm.

TUESDAY – MAY 23/16
Same work as yesterday, also new firing line work. We have
advanced a portion and it is not yet all joined up. We are doing this
and completing the unfinished portions.

This firing line work has to be done at night time only and we
have to have a covering party out front. I placed the covering party
my self and put it out 50 yds. We started work about 4 o'c and
stopped at one.

Weather: Fine and warm.

WEDNESDAY – MAY 24/16
Same work as yesterday and at night the firing line work. It was raining most of the day and night so made work very hard to do. The
c.o. returned during last night or early this a.m. and came to our
Company H.Q. in the afternoon, and asked me to retain the Command of the Coy until we moved out of the Trenches, so as to give
Capt. Ayre time to pick things up.

Weather: Rain part of day and most of night.

THURSDAY – MAY 25/16
As yesterday with regard to day work.

At night we had the same firing party in the line. As usual I placed the Covering party myself, as, being O.C. the Coy. I was responsible for the party. There are three other Officers in the Coy. Lt. K[eegan]. 2/Lt. A[yre],[33] and Capt. [E.] A[yre], and they go with this working party in turn; I also spend most of my time there too, partly owing to the danger of the place, as there is no wire in front, and partly to see how things are going, because I have to render a Work Report for each day and night at 3:30 a.m. daily (I also have 4 other reports to make at the same time) so want to get as much as possible done.

At 1 o'c a.m. when it was time to quit work I was out near the covering party when suddenly a flare was sent up and by its light I saw 3 Germans on the grass, kneeling, about 30 yds from me and 20 yds from one of my covering party. It seems that B. Coy's covering party, which was on the left of ours, saw that they were there, and Capt Nunns who was with the party at the time, took his Covering party in and did not know how to let me know they were there, so decided to send up a couple of flares to let me see them and I did. I then went to each man (12) and told him of it and to remain where they were until I told them to move in. I then sent in word for the working party to move in from the trench as we were just stopping work. And after 10 mins. I moved the Covering Party in, and nothing happened. The Germans were simply out to see what they could see. It would have been folly for us to have fired on them for immediately we would have done so, we would have had a dozen Machine Guns turned on us and the working party.

Weather: Rain part of day and night.

FRIDAY – MAY 26/16
Working as usual today. At night went out to continue work in Firing line, but it was very wet, so brought the men in after an hour or so
Weather: Wet all day and part of night.

SATURDAY – MAY 27/16
Working parties as usual today. I got orders to proceed to Mailly Maillet today to prepare billets for the Battalion to move into tomorrow.

I got back to the Trenches again about 6 p.m. Tonight we had to put up a Barbed Wire Entanglement in front of the New Line we have been making. Each Company had a proportionate share. I wanted D. Co. to be finished first, so had all plans made and the various steps systemized, and all N.C.O.s and some men shown the system we were to work on. We also had about 10 yards set up in our Company

Dump and the process shown and explained to all who were going on the work. We started at 9 o'c – first putting out our Covering Party. We all worked energetically and more or less silently, and we had the pleasure of being first finished; accomplishing our work by 12:30 – A Co. was second and B. Co. third and last.

Weather: Fine and warm, though a little damp.

SUNDAY – MAY 28/16

Today was spent in making preparations for moving out. Each Company is moving out independently by Platoons, D Co. starting at 3:30 p.m. I went out at 2 p.m. and took with me a guide from each Company. These guides were to meet their own Coy. at the entrance to Mailley Maillet, and show it to its billets, having been shown the billets by me previously, hence leaving at 2 p.m.

The whole Battalion did not get into its billets until about 10 p.m., parts of the relieving Battalion having been late.

Weather: A perfect Summers day.

EXTRACT FROM LETTER DATED 28 MAY 1916

I suppose you have been wondering why you have not heard from me for a fortnight. Well, I have not had a moment to myself since we went into the Firing Line ten days ago and we have just come out this evening. We had only about three day's rain during our sojourn in the trenches this time.

We have been furnishing fatigue parties, and doing various important jobs day and night. All men had to work four hours by day and four hours by night. Half the Company worked in the morning and half in the afternoon, then the whole Company worked at night from 9 p.m. to 1 or 1.30 a.m. They had to 'fall in' about 7.45 p.m.; when shovels and picks were issued, and the night's work explained by the Officer Commanding Company. They then marched in single file through the trenches to the place of work in the Firing Line. By the time we returned it would be about 2 o'clock, then 'stand to' from 2.30 to 3.30 a.m., which meant that it was 4 o'clock when we got to bed. I got only three hours sleep per night during this period.

Capt. [E.] Ayre, Lieut. Patterson and Lieut. W. Ayre took turns each night in going into the Firing Line, but I went every night, owing to the danger of the work and its importance and urgency, and being responsible for the welfare of the Company and the quantity of work done. Of course, each night we had to put out a covering party of twelve to fifteen men as we were working only two hundred and fifty yards from the German lines. Our Covering Party numbered fifty.

Upon returning from the Firing Line each night I had five reports to make out, namely, Situation Report, Intelligence and Operation Report, Ammunition Report, Casualty Report and the Company Work Report for the twenty-four hours previous to 4 a.m. that morning, and send same to Orderly Rooms immediately.

However, the main thing is that everything went off without a hitch, and we moved out of the Firing Line this evening, not having has a single casualty.

MONDAY – MAY 29/16

We had the usual fatigues that we get whilst in the vicinity, some of the men working by day and some by night. About half of "D" Coy. was on Special Town Duties, such as: Brigade Guard, Battle Guard, Aerial Observation Post, No. 1 Patrol, No. 2 patrol, etc., etc. The balance of the Company did mostly night work. I passed the Company over to Eric Ayre this morning, or rather last night to be exact.

I was engaged all today on further work in connection with the billeting.

Weather: Fine and warm.

TUESDAY – MAY 30/16

The Company did some day work today and some night work.

I was engaged setting up Englebelmer and Acheux [location of 29th Division HQ] billeting accounts.

Weather: Fine and warm.

P.S. I had about 16 miles horse-back ride today and thoroughly enjoyed it. I am still using the big horse which Col. Burton used to have. He used to be a very uncomfortable horse to ride, being very jumpy and falling on his feet with heavy thuds, but is now getting better and I am getting used to him.

WEDNESDAY – MAY 31/16

Our Coy. was again on night work, having to go to the support line just a little to the left of where we just moved out from. We were improving Communication Trenches, and making the Support line into a good firing line in case needed. I went out with 50 or 60 men tonight at 7:30 p.m. and returned about 2:30 p.m.

Weather: Fine and warm but slightly cool at night.

THURSDAY – JUNE 1/16

Today was spent as usual, in fatigue parties day and night.

My time was occupied in inspecting and allotting Cellars for the different Companies, to be used in case of a severe bombardment of the town by the Germans.

Weather: Fine and warm.

FRIDAY – JUNE 2/16

Company working parties day and night. I was out with the night party and we had several narrow escapes thro' being shelled just after starting work. About a dozen shells were sent over – three at a time – and one time was on the firing line when one lot came over, and as they were fired at very short range, the shell arrived at the same time as the sound, with the result that I hardly had time to take cover; however, I managed to duck below the parapet and glancing up saw the flame on the edge of the parapet. I can hardly tell how our men escaped unharmed for they were scattered all thro' the trench, a couple of Essex next to us, within a few yards, were wounded, but it was one more instance of our luck.

Weather: Fine and warm by day but a little cool at night.

SATURDAY – JUNE 3/16

Working parties as usual today but none at night. This was on account of our Batteries' intention of giving the Germans a good Bombardment in honour of the King's Birthday.

The Bombardment started at 12 o'c Midnight and did not finish until nearly 1:30 a.m. It was certainly a very heavy bombardment, about the heaviest we have seen or heard. It was a dark night and rather cloudy and so the flashes showed up the more, in fact there was practically a continuity of flashes, so vivid that the yard of our billet was made as bright as day, and one could plainly see everything in the yard, whilst the reports were one continuous roar. All windows, doors, and even chairs etc., were kept continuously rattling, in fact everything movable in the room was kept "on the shake". We used *some* of our really big guns (though not the biggest) and the reports of the explosions from them was really terrific, and was certainly awe-inspiring.

The Germans replied quite reasonably, too, but none of their shells dropped in our village, Mailly Maillet. Am glad we were not in the firing or Support lines for this.

Weather: Fine though dull.

SUNDAY – JUNE 4/16

Everything was as quiet as usual today and, being Sunday there was less work for the men.

Weather: Fine in morning, with rain in the afternoon.

P.S. On Friday I omitted to mention that I had a very enjoyable horseback ride, thro' Beausart, Berthamcourt, Louvencourt, Acheux and back to Mailly Maillet – 16 or 18 miles. At Louvencourt I met an officer whom I knew at the Peninsula and he invited me to Tea so had Tea with their C.O. – R.D. [Royal Dublin] Fusiliers – Country beautiful – all green.

MONDAY – JUNE 5/16

Working parties as usual today and night. I was out again tonight from 7:30 p.m. to 2:30 a.m.

Weather: Fine and warm – cool at night.

TUESDAY – JUNE 6/16

The Battalion was working as usual again today. I spent the day in squaring up our Billeting Account in Mailly-Maillet as we were moving to Louvencourt tomorrow. In the afternoon about 3 o'c I went on my horse to Louvencourt (7 or 8 miles) to prepare billets for the Battalion. Whilst there I heard that Lt. Col. Kelly [commanding officer] of the Essex, and previously of the S.W.B. [South Wales Borderers], who had been wounded on the night of the bombardment – June 3rd – had since died. He had a great name for daring recklessness, and got his wound by carelessness and foolhardiness; he was sitting upon the parapet watching the effect of the bombardment, and anyone who did such a thing deserved what he got.

I arranged some of the billeting and got back to Mailly-Maillet at 9 p.m.

Weather: Raining heavily all morning and a little in afternoon, clearing in evening.

WEDNESDAY – JUNE 7/16

I went to Louvencourt again this morning and completed billeting arrangements. Whilst there I was thunderstruck to hear that the H.M.S. "*Hampshire*" with Lord Kitchener with part of his Staff on board, had been lost with practically all hands. It seems too bad that Kitchener should lose his life in this manner. Then of course this will give the Huns a good deal of encouragement, tho' it will not affect us in any way, and will tend to postpone the day of Peace.

About 2:30 p.m. some of the Batn. details arrived and the first of the Batn. put in appearance at about 4 p.m. As part of "C" Coy. did not arrive until 8 o'c or later it was about 11 o'c p.m. when I went to my Billet.

Weather: Moderately fine – little rain.

THURSDAY – JUNE 8/16
I was occupied today in looking over and finalizing the Billeting. The Battalion was doing a practice attack, – preliminary training for our great push.
Weather: Fine – but dull at times.

FRIDAY – JUNE 9/16
This morning I commenced a course in Bombing as did also Capt. Butler. The C.O. sent for me yesterday and told me he wanted me to take over the Bombers and instruct and practice them, but I said I did not feel qualified to do so and would like to be able to take a course first, hence the above. It looks now as if Butler is going to be Bombing officer and I a reserve. The course we are attending is being conducted by the Brigade B.O. and consists of about 20 in all. We have a Theory Class in the morning and Practical in the afternoon.
The Battalion was having a Tactical Practice again today.
Weather: One or two small showers.

SATURDAY – JUNE 10/16
Bombing Course for me again today. The Battalion fell in to march off to another Tactical Exercise but it commenced to rain very heavily so it was called off, but an hour later it cleared so they fell in and set off but whilst they were away it rained worse than before they started, so when they returned around 1 p.m. all were wet.
Weather: *Raining most* of the day.

SUNDAY – JUNE 11/16
Am now acting Second-in-Command of our Coy. Went to Church at Brigade Headquarters at 11 a.m. (though not detailed) where there were assembled about 3000 troops. The Divisional Chaplain was the preacher and he made a touching reference to the Late Lord Kitchener.
After dinner the whole of our Coy. was detailed for work at Acheux, we being duty Battalion today. Parts of the other Coys. were out this morning. At 3 p.m. I went to Acheux with 100 men from B and C Coys. and when we got there found there was no work for us to do. By the time I phoned a message to the Adjutant, got a reply and marched back to Louvencourt it was 7 p.m. The C.O. was very vexed that we had been to Acheux and wasted the whole afternoon for nothing, so reported the affair to the Brigadier General.
We heard today and it has been rumoured the past couple of days, that Col. Kelly is not dead and is now getting on fine.

There are many big guns passing thro' here daily – a dozen yester-day – and half a dozen Howitzers today two 8 inch and two 9.2 inch.

Weather: *Fine and warm*, but cool in evening.

MONDAY – JUNE 12/16
The Battalion was out on its Tactical Exercise again today. The idea was this: A large area of ground equal in extent almost to that over which we are to attack shortly in what will be without doubt, the greatest in the history of the world, and about similar in aspect and contour. All the various German Trenches, roads, etc. are also marked with flags. We go over the ground exactly as we will on the great day.

I was still on Bombing today.

Weather: Rainy at periods.

EXTRACT FROM LETTER DATED 12 JUNE 1916
Was unable to write my weekly letter yesterday, for though we are supposed to be out of the Firing Line for a rest, it is only a change, for we are really working harder, though. we do have the advantage of getting a full night's rest. In the ordinary course of events. We would have been free yesterday, being Sunday, but it was our turn to be Duty Battalion, so we were busily engaged on all kinds of work.

You asked if I meant for you to send no clothing and no eatables as well. Well, we require no clothing, and if we do, Jim can get his now from the Regiment, and I can send to England for mine. As regards cake, chocolate, &c. the expense of sending it so far is really out of proportion to the value. However, twenty-seven bags of mail, all parcels, arrived for our Regiment a few days ago, and Jim and I had the pleasure of receiving the parcels you sent each of us, for which we thank you very much indeed. They were, of course, in good condition, in fact, one would not expect otherwise when such care in its packing has been taken.

We were all very much surprised to hear, a few days ago, of the big Naval Battle [Jutland] which took place in the North Sea. We have not yet received full particulars, but at present understand that the Germans lost fifteen ships, and we, the large number of fourteen. It is surprising that we have to learn everything by bitter experience and always to our cost and generally a heavy one. The loss of life must have indeed been heavy.

TUESDAY – JUNE 13/16
The Battalion was again on its tactical work, and I on Bombing.

Weather: Rain at intervals.

WEDNESDAY – JUNE 14/16
My Bombing course being now over, I went with the Battalion on Tactical Exercise.
   Weather: Rain nearly all day.

THURSDAY – JUNE 15/16
We all got up at 5 a.m. today – all clocks and watches were, by Army order, put on one hour last night at 11 p.m. which shortened our sleep, apart from getting up early. The Battalion left at 6 a.m. for another tactical exercise. The reason for going so early was that we were going into the Trenches again in the afternoon. I asked the c.o.'s permission to remain behind and attend to the squaring up of our Billeting Acct. here at Louvencourt. The Battn. returned at 11 a.m. At noon I started off for Englebelmer to prepare billets for the Quartermaster's Dept. and also for 50 men and three Officers who were going there to practise for a raid on the German Trenches. As Capt. Butler is amongst them the c.o. informed me that he wanted me to take over his duties as Intelligence Officer and to take over charge of the Snipers as well; this is besides being Bombing Officer.
   The Battalion moved into the Trenches after 5 p.m. They were very muddy after all the recent rain, but fortunately for us the rain has stopped and there is every appearance of a change.
   Weather: Dull but no rain.

FRIDAY – JUNE 16/16
Today the Germans sent over 50 or more shells of all sizes and were apparently registering. They did a little damage to our trenches and five men were slightly wounded.
   Weather: Fine and warm but quite cool at night.

SATURDAY – JUNE 17/16
The Enemy's Artillery was fairly active all day and it was apparent they were doing so for registration purposes. We had two men slightly wounded. Six German Planes was seen together over our lines but well to our left, and three at another time. They were soon sent away by our guns.
   Great preparations are now being made for the great push, which comes off next week.
   Our Division, 29th, is taking a Brigade frontage. The 86th is to take the first line, which will be a cinch as the first line is sure to be empty, then the 87th is to take the 2nd line and the 88th, our Brigade, is to take the 3rd line, which is certain to be the hardest, besides which we have to walk 5000 yds to it and will be under shell

and Machine Gun fire all the way. The Worcesters and the Essex in Artillery formation were to go first, and ourselves and the Hants behind them, respectively.

Weather: Fine and warm – cold at night.

SUNDAY – JUNE 18/16

The Enemy Artillery was very active on our lines this morning from 8 a.m. to 1:30 p.m. sending over about 70 shells 10 – 77mm – 35 10.5 cm. and 25 15 cm. A bridge was smashed, parapets etc damaged and 2 men wounded.

One of our snipers caught a German using a telescope this morning and succeeded in getting a bullet into him.

Enemy Planes were unusually busy today – five were over our lines at 5 p.m. and 4 at 8 p.m., but were very high; they were eventually driven off.

About 8:30 p.m. we saw a nice little duel between one of our Biplanes and a German Biplane. The German Plane was very much faster than ours and was able to turn in half the time ours could, nevertheless it was our plane which did all the following. First it was long sweeps with a sharp turn, then round in circles and then off again. They used their machine guns a bit but it appeared the German was only playing with ours, for he used to get away easily then turn and come back again.

Recd. a Pork Pudding today from Livermead.

Weather: Fine and cool but cold at night.

MONDAY – JUNE 19/16

This was a very annoying day, for we have been subjected to very severe shelling all day. From 8 a.m. to 7 p.m. the Germans sent over about 500 shells 3 inch – 4.2 inch and 6 inch, a very large proportion of them being the 6 inch. Strange to say, we had only 3 casualties, all wounded – one being Munn[34] (W.A.'s son), but parapets, communication trenches etc. were badly damaged in many places.

As I am now messing in H.Q. Mess, I took my Pork Pudding down for lunch (I am Mess Pres.). Needless to say the, the C.O. hurrahed when he sat down. Jim received one also.

Weather: Fine and cool, cold at night.

TUESDAY – JUNE 20/16

Today has been a moderately quiet day. The Germans' gifts to us today, in the way of shells, amounted to only 70, but strange to say we had four men wounded, one of whom may die, for the main artery of one of his legs has been severed; he is an Assyrian named

Joe Sheehan.[35] One of the men wounded yesterday, has since died, also a man who met with a bomb accident, a piece going into his eye and probably touching a portion of his brain. There seems to be a strange pensiveness about everything, and we are all strangely thoughtful about the "Great Push".

The bombardment was to have commenced today about daylight and was to last four days, and we were to mount the parapet on Saturday June 24th. This would have been a very strange combination of coincidences, for we are to a certain extent, from St. John's, at least St. John's is the Capital of our Island Home; we were to be lined up and start from a trench which we made ourselves when we first came to these trenches, and which has the official name St. John's Rd., and we were to move forward on St. John's Day – June 24th.

That is, "St. John's men, starting from St. John's Rd. on St. John's Day." However, for some reason or other, everything not being ready somewhere, affairs have been postponed for a few days, but we do not know for how long. The long and tiresome period of preparation and expectation is almost as bad as the real thing will be, but we are doing everything we can to be fully prepared for all contingencies and exigencies.

Weather: About the same as yesterday, though perhaps not quite as cold as last night.

WEDNESDAY – JUNE 21/16
The Hun certainly appears to be expecting our visit, for they are, according to reports all along the front, hard at work. There is an immense amount of traffic everywhere. Opposite our own particular position, he is seen working by day and night – with great care, of course, and there is quite a lot of traffic. Only last night, he could be plainly heard strengthening his wire work, and even adding to it, etc. Then his Aeroplanes are busy all day seeking information, by flying over our lines, but our A.A. [anti-aircraft] Guns and Planes, give them very little chance.

Weather: Very fine and warm – little cool at nt.

THURSDAY – JUNE 22/16
Things were unusually quiet in our line today. The Germans sent over less than 20 shells, but, no doubt, they are very busy in other ways. They cannot fail to see what is going on on our side of "No man's land", for we are as busy as we can possibly be by day (with extra care) and by night, building bridges across the tops of the various trenches, shrapnel and splinter proofs [corrugated iron sheets] to protect us from the enemy's reply when we bombard,

carrying water in Petrol Cans and storing in Reserve in Dugouts, Etc., Etc.

Aeroplanes have been very busy lately on both sides. There is a constant hum of their engines, from before daylight until after dusk; Sometimes one can see as many as 12 or 15 up, though they are nearly always all ours. Occasionally some of the Huns planes try to come across our lines, but are soon sent back by ours and by our A.A. Guns.

Four German Planes tried to come over this evening, but one of our planes flew right into their midst, and nearly did for a Hun. He made 3 or 4 complete spirals down towards the ground, but kept control of his machine and then flew off, whilst the others quickly made off. One of our planes was brought down by the Huns over opposite us today, being hit squarely by a German A.A. [anti-aircraft] Gun. Of course we are bringing down Hun planes frequently in one part or other of our lines. One was brought down a few miles from here a day or two ago – and we got her Logbook.

Weather: A perfect Summer day, very warm indeed and quite warm at night too.

FRIDAY – JUNE 23/16

Today is "moving day". I left the Lines at 9 a.m. to go to Louvencourt to prepare billets, having made arrangements for my horse to be at Auchon-Villers. I had to let the horse walk the whole 7 or 8 miles, owing to my knee being so sore. When going to the trenches on the 15th inst[ant], my horse "shied" at a dead horse lying in the roadway – it had been dead only a short time. When he shied there was a string of limbered wagons moving along the opposite side of the road in the opposite direction, the result being that my knee was brought with the force of the rapidly moving horse behind it against the wheel of a wagon coming towards me, the wheel band of iron is about 4 inches wide and half an inch thick. The blow was terrific and it paralyzed every muscle and nerve in my body, and split my breeches and drawers for about 4 inches and skinned my knee. This was nothing compared to what damage was done inside my knee. From where I left my horse I had to walk about 3 ½ miles to the trenches and by the time I arrived in I could barely walk, and it was swollen to fully twice its natural size. I showed it to the Doctor, who said he would see it again in the morning, and when he did he told me I would have to go to hospital, but I told him it would be alright in a day or two, so he agreed to give me 24 hrs before deciding to send me. In the meantime I kept constant cold bandages to it; the pain was very severe and when the next morning came it was no better, so he

(the Doc) again said I would have to go, and also told Col. Hadow.
But I asked the Doctor to give me another 24 hrs. to which he
agreed. By that time, am glad to say, the knee was somewhat better,
which saved me from hospital. I would not go to hospital seeing that
we have been preparing for "the Great Push" for a long while, and it
is at last just about to come off, besides which everyone would certainly have said I was "slinging the lead". Even now, a week after my
accident, I cannot help walking lame and only hope it will [be] sufficiently improved by the "great day" to enable me to do justice to my
duties. However, I eventually reached Louvencourt and spent the day
at the Billeting.

The ride down was quite enjoyable for it was quite a fine day. It
was certainly a wonderful sight on the way across country to see all
the immense preparations being made for the great coming event.
Every road in every direction was clouded with dust, being crowded
with moving transport wagons, ration carts, limbered wagons,
Troops in large and small parties, Etc., Etc. – And everyone seems so
cool about it all, quietly preparing for what is going to be the greatest attack in the history of the world, and very probably the greatest
there will ever be. We only hope that it may be a very strong factor in
bringing an early end to the war. Well, the battalion started to arrive
by platoons about 10 p.m. or 10:30 p.m. and it was 3 p.m. when
H.Q. arrived. The Signallers and Lewis Gunners were still not in and I
believe they arrived about 5 p.m. It was 3:30 a.m. when I had my
dinner, having had nothing to eat since midday.

Weather: Fine in morning, but a tremendous downpour for an
hour at 1 o'c – Never saw the like before – Showery for the rest of
the day.

SATURDAY – JUNE 24/16
Today was allowed as a day of rest to the men, on account of hard
time yesterday and late arrival last night.

I was busy finalizing the billeting.

The 5 days bombardment was to have commenced this morning,
but we have heard very little of it. This is because they are as yet only
using the small guns – but wait until the big ones start. – The 15 inch
ones have not fired a single shot since arrival. Met Lt. Col. Franklin
today and spoke with him for ½ hr.

Weather: Fine and cool.

SUNDAY – JUNE 25/16
Two German planes passed over here – Louvencourt today. Our A.A.
guns from vicinity of Reserve Line furiously shelled them, and a good

many pieces of the shrapnel, nosecaps, pieces of shell casing, etc., fell in Louvencourt. I was standing in the Garden of my billet and several pieces, some quite large, fell within two or three yards of me, so I was not long in getting into the doorway. One man in Louvencourt was killed by a piece of shell casing striking him on the head. Our Battalion was on Tactical Exercise.

Weather: Generally fine with occasional showers that soon passed away.

EXTRACT FROM LETTER DATED 25 JUNE 1916
The days are certainly flying by, for Sunday is here once more – and it is a fortnight since I wrote. What is more, I do not know whether this will go now or be held for some time, for I heard a week ago that all mails outward had been stopped.

We moved out of the trenches the night before last, and shall be going in again in a couple of days. This letter will be very short for we have been up to our eyes in work for the past fortnight and will be for some time to come, the reason for which I cannot give you even an inkling; but you will have heard all before this reaches you. Am afraid that quite a long time will elapse before you get another letter from us, but you need not worry for we shall be all right. Jim and I are both O.K. and in the best of spirits, and I hope it will *not* be long before we shall be *able* to write you a decent letter *each*.

Are not the Russians doing fine in the East. *Look out for us*. Am hoping that there will be a possibility of our spending Xmas with you all this year. Would not that be grand.

We are having quite good weather now and hope it will continue. We cannot give any news and have been specially told to speak or rather write of nothing but personal and family matters.

MONDAY – JUNE 26/16
At 9 a.m. we had a Battalion parade for the purpose of being addressed by Genl. de Lisle – G.O.C. [General Officer Commanding] of our (29) Division. He spoke for about 10 mins. He said he was glad to have the honour of addressing us as a battalion, for the first time, on the eve of what is going to be the greatest battle in the history of the world. He impressed upon us the dangers we would have to encounter, but that he had no doubt that we, who had the sole honour of representing Nfld., would bring honour and credit to ourselves and Nfld., judging by what he said and known of us in the past and that he knew he would have the pleasure in a very short time of sending a message to our friends in Nfld. telling them of the glorious work we had performed in this Great Attack. He also said that in the

4th Army front we had an overwhelming majority of guns and men, and an almost unlimited supply of ammunition (45,000 tons will be fired) (1 ½ million rds) for the guns. If our gun ammunition were loaded on trucks it would form a line of trucks 46 miles long. In men we had 21 Divns. or 263 Battns., whereas opposite to our Army Front the Germans had only 32 Battalions[36] and the most that they could bring up in a week was 65 Batns and the French have 20 Divns. on our right so we need fear nothing. When leaving he wished us Good Luck and Good bye.

The Battalion then spent the morning in having deficiencies in kit and new equipment required, issued. In the afternoon another Tactical Exercise was on, but when they reached the training ground it came on too wet so they returned. The Raiding Party is going out to raid the German Trenches tonight.

Weather: Rainy most of the day, clearing for short periods; fairly warm.

TUESDAY – JUNE 27/16
The Raiding party consisting of 3 officers – Capt. Butler, Lts. Strong and Green, and 57 men attempted their raid last night. They took a telephone with them and had connection with our Firing Line. They placed their 1st Bangalore Torpedoe in the wire and it did its work but there was still a lot of wire left, so they placed their second Torpedoe , but unfortunately it failed to go off, so they set their wire-cutters to work, but before they had finished, a few Germans commenced firing on them. As there was now only a very short time left before the artillery would commence shelling the German front line, they decided to abandon the attempt for the time being. However they are going out again tonight. Two men were slightly wounded – just scratches on the face, and were probably caused by barbed wire only.

The Battalion is on a Brigade Tactical Exercise this morning.

The raiding party went out again tonight. We hear practically no shelling going on but at night the sky is continually alight with flashes, but of course we are 8 miles from the firing line.

Weather: Dull and rainy.

WEDNESDAY – JUNE 28/16
Last nights' raiding party was more successful. When they got to the wire they found a great gap made by shell holes, so at once proceeded to go through. They were hardly thro' when they were fired upon from both flanks by rifles first and then Machine Guns, and it looked as if the Germans were fully prepared and waiting for them. Later they even

used Trench Mortars[37] on them. Three or four of the men got up close to the parapet and a couple got in. Bomb throwing was mostly resorted to by both sides, and an hour was spent in this miniature battle. Lts. Strong and Green were both wounded and went to hospital, but will both be O.K. after awhile. Capt. Butler got a slight knock over the right eye but will be alright in a day or two. Apart from the above our casualties were: ... killed, ... wounded, ... missing (believed killed).[38] However, several bits of useful information were secured, but unfortunately our men got no prisoners. Two or three Germans were bayoneted and many were sure to have been killed by the bombs [grenades] etc.

We were busy all day today getting ready to go off to the firing line tonight to commence the greatest attack in the world's history, tomorrow. All our things were packed and turned in to the Q.M. [Quartermaster] Stores, so as to be brot. on after us, and two or three hours before we were due to start, word came in that the move was off for 48 hours. We do not know the real reason, but believe it is owing to the weather, which has been very wet for the past week and still, and the prospects are that it will be so for a day or two. However, this will give us chance to have a further rest.

Weather: Wet and Rainy.

THURSDAY – JUNE 29/16

We were doing very little today. In the afternoon the battalion went out for a little training exercise just to keep the men fit.

Weather: Wet and Rainy.

FRIDAY – JUNE 30/16

Today was spent in getting everything ready to move tonight. Everything seems now in our favour in the weather line and we only hope that the artillery has done its share effectively, though we have heard very little, but I expect they are doing their work allright.

We moved off at 9:15 p.m. and had over 3 hours marching ahead of us.

Weather: Clearing during morning, and quite fine in the afternoon and at night, with the promise of a fine day tomorrow.

P.S. It is great to see how happy and light-hearted everyone is, and yet this is undoubtedly the last day for a good many. The various Battalions marched off whistling and singing, and it was a great sight. Of course this is certainly the best way to take things and hope for the best.

EXTRACT FROM LETTER DATED 30 JUNE 1916

Just one more last note before moving off. For certain reasons, a delay has enabled me to write this as my last note, for a short while at least, whereas, I thought my last – a few days ago, – would have been the last for a few days to come.

Re the soil of Belgium: The name of the injection, which you could not remember, is, I think, "Tetanic."[39] Of course, we all have our 'Field-Dressings' with us, which contain, besides bandages, a little phial of 'Iodine,' this being a disinfectant of the best kind.

The loss of Kitchener and his Staff was certainly a great blow to all; however, I believe the climax of our troubles will be reached within the next few days, (after which the day of peace will quickly draw near), though they will undoubtedly bring trouble to many. Jim and I are in the best of health and spirits, and I trust we may remain so. – This will be my last letter for a short while.

SATURDAY – JULY 1ST/16
[Typescript entry appended to diary]
To the Adjutant:-
Herewith, as requested, are the names of men of D Company, who went over parapet on July 1st, and returned unwounded.

1775	Pte.	V. O'Donnell
1167	"	H. Noseworthy
1484	"	J. Day
1972	"	R. LeGrow
1645	"	T. Fitzgerald
1059	"	F. Riggs
990	"	C. Taylor
1523	"	H. Baldwin
1581	"	W. Morris
1012	"	A. Meadus
1418	"	J. Penny
	"	A. J. Stacey
	"	E. Moores
998	"	A. Sullivan
1660	"	W. Day
1149	"	J. Frost

CHAPTER EIGHT

The Battle of Beaumont Hamel

(1 July 1916)

I n April the Regiment received the first of two drafts of reinforcements, the second arriving on 30 June 30th, on the eve of battle,[1] too
late to benefit from the month of careful training that preceded the July
first offensive. In the end it mattered little. The task eventually assigned
to the 88th on that fateful day was not the role for which it had
trained.

The plans for the offensive were supposed to carry Rawlinson's
Fourth Army a mile and a half across a fourteen mile front driving the
enemy from his first and second positions. In the Beaumont Hamel
area this meant an attack on the German Intermediate Line on the
Beaucourt Road, some 5,000 yards distant, an ambitious and deadly
challenge for 8th Corps and its forward divisions, the 4th and 29th.
The 86th and 87th Brigades were to lead off the advance securing the
First Line, while the 88th, with its leading battalions, the Essex and the
Newfoundland Regiment, would advance steadily under artillery barrage towards the German positions on the Beaucourt Road. The
attacking battalions were to deploy in a succession of assaulting waves
in the belief that each wave added impetus to the preceding.[2]

An ambitious artillery barrage for several days preceding the offensive was expected to cut the enemy wire and strike at his defensive
positions. Great reliance was placed on the effectiveness of this pre-

liminary bombardment. In many sectors, however, particularly at Beaumont Hamel, on the day of the battle it was discovered that the enemy wire had survived the bombardment and some German trenches were untouched. More serious, this preliminary bombardment had alerted the enemy that an attack was imminent.[3]

Bad weather postponed the offensive until 1 July, allowing a few days of extra time for rest, reflection, and final preparations, including readying the sixty-plus pounds of personal equipment that each soldier was to carry into battle. At 9 P.M. on the 30th the regiment turned out for the final time: 25 officers, 776 N.C.O.s and other ranks.[4] Among them, young men in their twenties and early thirties, fishermen, lumbermen, trappers, and office workers, most veterans of Gallipoli and Egypt, others novices just arrived on the front. Steele wrote in his diary on the eve of the battle: "We were busy all day getting ready to go off to the firing line tonight to commence the greatest attack in the world's history."[5] On the final night soldiers huddled in the trenches warding off the damp and cold, some enjoying a quiet smoke or chat, others dozing fitfully, wearied from the long day's march. Last-minute inspections preoccupied the officers as they readied their commands for the long-awaited offensive the next morning.[6] For the young men who waited in the trenches that night, this was to be their first experience of going over the top. There was, no doubt, some sense of foreboding, as there must be before any battle, and yet from all accounts there was little evidence of last-minute jitters. "It is surprising to see how happy and light-hearted everyone is," Steele noted, "and yet this is undoubtedly the last day for a good many. The various Battalions marched off whistling and singing, and it was a great sight."[7] Most were confident that they were well prepared for what lay ahead and that the day of battle would mark the approach of peace. Steele's last entry in his diary expressed his own optimism of the outcome: "The climax of our troubles will be reached within the next few days after which the day of peace will quickly draw near."[8] It was a view shared by his brother Jim.[9]

Promptly at 6:25 a.m. on the first of July the artillery bombardment began and an officer from each battalion hastened to brigade [headquarters] to synchronize watches.[10] At 7:20 a.m. the mines under Hawthorn Ridge opposite Beaumont Hamel were set off in a tremendous explosion. Two platoons of Royal Fusiliers immediately rushed forward to occupy the high ground, only to be met by devastating fire from the outer lip of the crater. This was followed by an equally unexpected and withering barrage on the British trenches and across No Man's Land.[11] Almost from the outset the 29th Division's attack began to falter. First the 87th and then the 86th Brigade succumbed to the weight of enemy fire.[12] Confusion was compounded by poor communications – white flares sent up by the Germans were mistakenly taken by General de Lisle as a signal that the 87th Brigade had reached the first objective. From the support trenches the Newfoundlanders were unaware of the fate of the advancing columns and yet had begun to sense that things were not quite as they should be.[13] An hour and a half later the 88th was ordered to move with the Essex and Newfoundland Regiments advancing "as soon as possible" in support of the attack supposedly taking place on the divisional right. When Hadow asked Brigade Headquarters whether "as soon as possible" meant independently of the Essex, he was told to move as soon as he could. Hadow called a hasty conference with his company commanders but there was no time to change tactics so at 9:15 a.m. the message went to Brigade Headquarters – "Newfoundland Regiment was moving off." The Newfoundlanders were heading off completely on their own. The regiment's objective was to occupy the enemy's first trench system, a line of trenches south of the Hawthorn Redoubt on the edge of Y Ravine, and then to work forward to Station Road, clearing the enemy trenches there, some 650 to 900 yards distant.

In magnificent order, as practised so many times before, the Newfoundlanders moved out of their trenches down the exposed forward slope towards No Man's Land, the rear sections waiting until those forward reached the required forty yard distance ahead of them. No friendly artillery fire covered their forward movement. The advancing waves approached the first line of their own wire under heavy machine gun fire, first from the right and then from the whole German front. From their first line of trenches on the fork of the ravine, the Germans were perfectly positioned to inflict a murderous cross-fire on the advancing Newfoundlanders. Men began to drop, although not at first in large numbers. Only when they reached the first gaps in their own wire did they began to fall in larger numbers. By this point Jim Steele had already taken over from his platoon officer, Lieutenant W. Ayre, who had been knocked down just outside their own wire. Doggedly, the survivors continued to move on. "The only visible sign that the men knew they were under this terrific fire," wrote one observer, "was that they all instinctively tucked their chins into an advanced shoulder as they had so often done when fighting their way home against a blizzard in some little outport in far off Newfoundland."[14] As the remnants of the leading two companies emerged from their own front wire, they could see that the German wire was still intact. At this point Jim Steele was knocked unconscious by a wound to the head. Some determined souls made it through the inferno of fire to the German lines, only to be cut down as they tried to cut a passage with their clippers.[15]

The battalion headquarters, consisting of Hadow, Raley, a signal-man, several stretcher bearers and two runners, moved to a forward position after the advance began. In the fog of battle confusion reigned. Anthony Stacey, one of the runners, described the unfolding scene this way:

I could see no one moving, but heaps of Khaki slumped on the ground. Raley saw a man crawling into our trench on the left and said to me, 'Stacey, go along the trench and see what you can find out from him.' I started to push my way through the trench ... but it was filled with the wounded, and as I tried to push by the wounded, agonizing cries would come from them. It was too much for me so I turned and went back and told Raley it was impossible. We stayed there for a while because it was useless to go forward as it seemed that everyone was wiped out.[16]

In less than thirty minutes it was all over. At 9:45 Colonel Hadow, who had witnessed the annihilation of his regiment from a forward position, made his way back the 100 yards to Brigade Headquarters where he reported to Brigadier Cayley that the attack had failed. Incredibly, Cayley ordered Hadow to collect any unwounded New-foundlanders and make a renewed attack on the German trenches but he was unable to find any Newfoundlanders among the dead and wounded that filled the trenches. Fortunately, the order for a suicidal attempt on the German lines was countermanded by a senior officer from Division.[17]

Fearing an enemy counter-attack, Divisional Headquarters issued orders to consolidate the line, which brought the regiment's reserves to the front. Several of the reserves were already wounded. Through-out the day survivors of the morning's offensive had attempted the long and dangerous journey back to their own lines, many becoming the targets of German sharpshooters or the victims of the enemy's constant bombardment.[18] Ron Dunn[19] lay wounded on the battlefield for several days. On the second day, convinced that he would soon die, his thoughts turned homeward to Bonavista Bay and his mother. "I said me prayers," he recalled and then drifted off, unaware that rescue was on the way.[20] Jim Steele waited patiently until dark to attempt a return to the regimental lines. That evening, for an hour, divisional artillery fired on the German lines to cover efforts to recover the wounded. When roll call of the survivors was taken only sixty-eight responded.[21] The full cost would not be known several days. The final grim figures revealed that the regiment had been virtually annihilated: 14 officers and 219 other ranks killed or died of wounds, 12 officers and 374 other ranks wounded, and 91 other ranks

The Newfoundland Regiment's objective, 1 July 1916. An official
photograph of Beaumont Hamel taken after its capture in November 1916.
With compliments from "The Veteran" Christmas 1928.

missing, a total of 700 all ranks, second only to the 10th West Yorks,
which suffered 710 casualties.[22] Most were probably struck down
before they even reached the British front line. The loss of officers
was staggering: every officer who went forward that morning was
either wounded or killed. Four members of the Ayre family were
killed on that day: Captain Charles Robert Ayre; his only brother,
Captain Bernard P. Ayre, who was serving with the Norfolk Regiment, near Maricourt; and two cousins, Lieutenant Gerald W. Ayre
and Lieutenant Wilfred D. Ayre, both officers with the Regiment.[23]
As Lt Col Hadow later recalled: "The extra-ordinary thing [was] how
anyone had survived it at all."[24]

As part of the mandatory ten percent reserve Owen Steele did not go
over the top that fateful morning but remained behind as Officer Commanding D Company. In the confusion that followed the carnage of
1 July, Steele must have had little time to think about his diary. His
brother Jim had survived the battle, which would have brought obvious relief as the two were very close. In one of those peculiar circumstances that so often occur in war, Steele was struck by a stray shell on
7 July while standing outside his company headquarters at Colin
Camps. Captain Forbes-Robertson was the first to reach him. Steele

Lieutenant James Steele

was alive at the time and seemingly not badly wounded. In fact, a piece of the shell had struck him in the thigh. The doctors tried to save him by amputating his leg but he died the next morning.[25] He was buried at Mailly Maillet Communal Cemetery Extension, about 9 kilometres north of Albert.

Jim was transfered to Wandsworth Hospital in England, where he heard of his brother's death. Surprisingly, his own rather sparse diary entries make no mention of Owen's death. Nor does his written recollection forwarded on to the Ministry of Militia four years later include any reference. One can only surmise that the memory was still too painful to be put on paper.

CHAPTER NINE

A Family Grieves – A Nation Mourns

For several days Newfoundland remained unaware of the terrible tragedy that had befallen its regiment. The initial reports of the Somme Offensive had been unduly optimistic. A jubilant Bonar Law, secretary of state for the colonies, cabled Governor Davidson late in the afternoon of the 1st with the news that "Our troops have broken the German forward defences on a sixteen mile front." [1] Five days later the full extent of what had actually happened began to filter down into the local press. Faced with the overwhelming number of casualties, the government dispensed with individual notifications. Family and friends, in consequence, poured frantically over the long lists of dead and wounded. "Today all Newfoundland is in sorrow" – the *Daily News* proclaimed in a solemn editorial on the seventh. Its eulogy was typical of the immediate public response: "The price of Freedom is paid in tears, and the maintenance of liberty is the blood of its defender ... our brave and noble boys have given freely and generously of themselves in the service of humanity." [2] St. John's was hardest hit. There was hardly a street in the capital where a loved one, a family member, relative, or friend killed at Beaumont Hamel had not lived. [3] "St. John's is weeping for her dead and wounded. What can comfort her? Even peace cannot dry her tears. Time alone may heal such wounds." [4]

It must have been a great comfort for the Steele family to learn that Jim and Owen had survived the battle. The unexpected shock came on the tenth when the family learned that Owen had died of wounds two days previously. The full details arrived several days later. The family was devastated. For Samuel the loss of his eldest son seemed a bitter irony: "It is not right that any one individual should expatiate on their own sorrow in these terrible times but it does seem hard that when he had gone through the first offensive and deserved the rest they were then taking, that a stray shell should have reached his billet."[5] Sarah Steele wanted her son's remains brought back home, a move that Governor Davidson wisely discouraged: "I need hardly point out how difficult it is in the midst of a desperate struggle to make provisions for the seemly removal from the scenes of battle of the remains of our gallant soldiers." There were also precedents, Davidson pointed out, going back to the Boer War that the remains of fallen soldiers would not be repatriated but left to lie in tended marked graves.[6] Instead he supported the family's effort to retrieve special souvenirs associated with Owen's service – his diary and his Brownie camera. For the parents these two items seemed to offer a personal link with their dead son, a diary written in his own hand recounting his experiences overseas and photos taken at various times, including images of their beloved Owen.[7] Over the course of the next few years the family received other items connected with Owen's service with the regiment – his medals, his commission as lieutenant in the imperial service, the regimental headquarters brigade flag. Their most important concern, however, was to memorialize their son's war service through the publication of his diary.

Steele approached Governor Davidson in the late autumn of 1916 with an offer to publish 1,000 copies of the diary, including photographs taken by Owen. The proceeds would go to the Newfoundland Patriotic Association. Davidson agreed to write a foreword.[8] His daughter Ella worked tirelessly to prepare a 250 page typescript of the diary, which was bound and presented to Davidson in December 1916. Wartime requirements meant that the diary and photographs had to be sent to the deputy chief censor for review,[9] who required erasures on thirty-four pages. These ranged from the names of transports that had taken the first Canadian Division to England and were still employed on transport services, to the names of places behind the Somme front where the Newfoundland Regiment had been billeted and trained and where British troops were still located, to details connected with the death or injury of regimental members which were thought to be upsetting to families. The changes were, in fact, minor.[10]

Sarah, Samuel, and Mollie Steele at Owen's grave at Mailly Maillet

By the time Davidson received the revised copy of the diary the War History Committee of the Newfoundland Patriotic Association had already prepared its own version of the colony's war effort, based in large measure on Steele's account. Their account was to be serialized in the local press. As Davidson admitted, the two accounts "must necessarily overlap" leaving him with a dilemma over how to deal with the Steele initiative.[11] In the end, nothing was done. The bound typescript of the diary was returned to the family, dashing all hope that Owen's death would be publicly acknowledged through publication of his wartime journal. One can well imagine the bitter disappointment for a family still in mourning. Ironically, the War History Committee's account was never published either. In 1921 Richard Cramm published the first account of the regiment's war history, *The First Five Hundred*, which included extracts from Steele's diary. The official history finally appeared some fifty years later, written by army historian G.W.L. Nicholson. It includes frequent references to Steele's diary.

The family went through a necessary but lengthy process of mourning. On the fourth anniversary of Beaumont Hamel they and others gathered for a special ceremony at Queen's Road Congregational Church to mark the unveiling of a tablet in memory of men of the congregation who had fallen in the Great War. There were thirteen names

Permanent grave marker on
Owen Steele's grave

on the tablet, including Owen W. Steele, a former choir member who
had also served on the church's board of managers The service began
with a passage from 2 Samuel 1:19, "The beauty of Israel is slain upon
thy high places: how are the mighty fallen." As the names of the dead
were read out the governor unveiled the memorial tablet. The service
followed the traditional forms of the Christian liturgy, holding promise
of salvation and eternal rest for those who died in the faith. For a religious family such as the Steeles, the memorial service must have afforded some degree of comfort.

Later that year Samuel and Sarah and their two eldest daughters,
Mollie and Beatrice, made the pilgrimage to Owen's grave at Mailly
Maillet. The faded photograph of the occasion shows the two grieving
parents and daughter Mollie looking down at what appears to be a
fresh grave site. The plain wooden cross driven into the mound of earth
covering the grave had a small brass plate with the simple inscription
"Lieut O.W. Steele." For Samuel and Sarah, the occasion was the ful-
filment of a personal mission, expressed soon after Owen's death, to
visit their son's grave once the war was over. Later, when a permanent
grave marker was erected on the site by the War Graves Commission,
the word "Sacrificed" was inscribed on its base, a personal request by
Samuel Steele.

The final stage in the grieving process came with S.O. Steele's attendance at the inauguration of Newfoundland Memorial Park and the

unveiling of the Newfoundland "Caribou" monument at Beaumont Hamel France by Field-Marshal Earl Haig on 7 June 1925. It was in every respect a stellar occasion with ceremonies that lasted over two hours. Keynote addresses were given by Earl Haig, Britain's most distinguished soldier; the Hon. J.R. Bennett, Newfoundland's colonial secretary; by the former commander of the 29th Division, Lt-General Sir Beauvoir de Lisle; and by Marshal Fayolle, chief of the French General Staff. Steele took only three photographs: one of General de Lisle, another of the inauguration by Haig, and a third of Bennett speaking with the magnificent caribou memorial as a backdrop. Shortly thereafter Samuel, his wife, Sarah, and three of their daughters left for England, never to return. It has been suggested by their grandson that the senior Steeles never recovered from the death of their beloved Owen. Perhaps the Newfoundland associations were too painful for elderly parents to bear and so they sought solace elsewhere.

For the nation, time helped to heal the wounds of war and blur the memories of the tragic cost of Beaumont Hamel. It was early June 1919 before the first draft of the regiment returned to its native soil to a tumultuous welcome.[12] On 1 July that year, the residents of St John's gathered for the first public commemoration in Newfoundland of Beaumont Hamel. In the years that followed Newfoundland erected permanent memorials to the dead of the Great War, some on the island, others on distant battlefields. In 1924 Newfoundland Memorial College, the predecessor of the present-day Memorial University of Newfoundland, was opened as a living memorial to the nation's war dead. That same year the National War Memorial was completed on a site overlooking the spot where Sir Humphrey Gilbert, the Elizabethan explorer, had laid claim to Newfoundland for Queen Elizabeth I in 1585. Field-Marshal Earl Haig unveiled the war memorial on 1 July 1924, eight years after the Battle of Beaumont Hamel and nearly ten years after the war had begun. In a moving ceremony set against the stark granite backdrop of the new war memorial Haig offered an eloquent tribute to the regiment's "high courage and unfailing resolution" and his heartfelt sympathy to those who had lost relatives in the service of their country.[13] The unveiling of the war memorial in St John's and a year later the memorial park at Beaumont Hamel allowed Newfoundlanders to formalize their grief and to acknowledge publicly those who had fallen in the service of their country.

We are now eighty-six years removed from Beaumont Hamel and yet Newfoundlanders still recall, on the Sunday closest to 1 July, at 11:00 a.m. precisely, the "Best of the Best," the young men who made the supreme sacrifice for King and Country.

"In those four years of strenuous war, a lifetime was crowded, whose sensa-
tions and experiences were more appealing and more appalling than those
which thread the average three score years and ten. As the years pass the ranks
grow less, but the memories more sacred. Some one of the Blue Puttees ... will
be the last survivor of the gallant lads whose patriotism and merriment made
Pleasantville ring and the hearts of Newfoundlanders swell with pride and
gratitude."[14]

Newfoundland's last link to its great tragedy,[15] the last of the Blue
Puttees – Abe Mullett, died in July 1993 at 101. A stretcher bearer
attached to the Red Cross, he witnessed the carnage of 1 July 1916 first
hand – retrieving the shattered remains of hundreds of his fellow
Newfoundlanders from the bloodied battlefield of Beaumont Hamel.
The last survivor of the battle, Walter Tobin, died in 1995. A popular
figure, he graced many an occasion connected with the regiment and
the Great War. Towards the end of his life, Tobin grew cynical about
the war and his own part in it, despairing of its misplaced idealism and
its waste of young lives. The "war to end all wars" had been the
seedbed not of peace but of future wars.

Acrostics and Poems

From letter of November 5, 1915

FIGHTING AFTER TRAINING HAS EARLY REWARD.

Since that far away day when to war we set forth
And left far behind us our dear Newfoundland,
Many paths have we travelled, from Scotland's cold North
Until we eventually reached Egypt's hot sand,
Ever determined to give expression to wrath,
Leaving nought of the Kaiser and his cruel German band.

Oh when will this struggle titanic be o'er,
When the nations of Europe will at peace be again,
Enabling us all to go homeward once more?
"Not till we prohibit the Kaiser to reign."

So let us fight on for "Der Tag" is quite near,
That surely will see this work properly done,
Ever and always, in whatever sphere,
Evade not our duty nor from. the fight run,
Let the world see that we do not know fear,
Even when faced by a big Turkish gun.

Acrostic Composed by O.W. S.
In the trenches – Gallipoli Peninsula – Nov. 5, 1915.

From letter of 5 November 1915

MAY OUR TRAINING HAVE EARLY REWARD.

Sometimes the beauties of the sky
At setting of the sun,
Remove all thoughts of battle-cry,
And silently, they make one sigh:
How can a war be on?

Big fluffy clouds, so fleecy white,
Lend beauty to the scene,
And then the clouds, in, shades so light,
Now Pink, and White, and Green,
Complete a most exquisite sight.
How is it that the Nations fight?
Each side has "Rights" to screen.

So fight we must, and fight we will,
Till, victory is ours.
Except we conquer "Kaiser Bill, "
Explode his mental towers,
Let's fight; our foes must have their fill
Ere Britain. rules the Powers.

Acrostic Composed by O.W. S.
In the trenches – Gallipoli Peninsula – Nov. 5, 1915.

Copy of Enclosure in Last Letter June 30, 1916

THE WOMAN'S PART.

The boys in their khaki go out to the front! What are the women to
    do?
They say, "Men must work, and women must weep." Is that all that's
    left for you?
Don't believe it! The hardest part to play is the part of the mothers
    and wives:
To give your own life is a little thing! We give our men-folk's lives.
The baby you've borne end suckled, and put in his shortend, frocks –
The boy that you've often scolded when you washed him and darned
    his socks:
We've bred them, and reared them, and loved them – and now it's the
    woman's part
To send them to die for England – with a smile and a breaking heart!

And we'll do it! Oh, girls might trifle, in the careless days of peace,
With the boys of the seaside bandstand, his flannels without a crease.
We might flirt, and kiss, and flutter – but the day of the war began
We women had done with the loafer – what we want to-day is a
    man!
The man that will shoulder a rifle, and go out where the bullets fly,
With his head held high, and a song on his lips, and a smile as he
    says "Good-bye."
We'll bid him God-speed and wish him good luck, and tell him he's
    one of the best,
And he'll soon come back, with his duty done, and the hero's cross
    on his breast.

There's no place for a girl in the fighting line – but let this be your
    woman's plan –
If we can't enlist for service, we can each of us send a man,
If he lags, wake him up with a scornful word! Let him feel the lash of
    shame!
Till you fire his soul to ardour, and kindle his blood to flame.
Let it be "Hands off" for the sluggard! For the knut and the flapper's
    joy,
No smile and no kiss for the shirker! Keep your lips for the soldier
    boy.
Send your boy to the colours, mother! Hand him his belt and gun –
It's better to lose him nobly then to be ashamed of your son.

When the work of the day is over, you can let yourself go, and cry
In the gloom of the desolate fireside – in the dark, when there's
    nobody – by.
There isn't a sock wants darning; there isn't a boy to scold
For the cigarette ash or. the carpet for the dinner they 've let grow
    cold.
Their caps still hang on the hatstand, but there isn't a step or. the
    stair;
There's no gay young voice calling "Mother!" No sound – for the
    boys aren't there.
That's the time you know the anguish of the waiting woman's part:`
In the hush of the lonesome home it's the silence that tears your
    heart.

Night passes! We'll welcome the morning with a smile and a steadfast
    will;
If we haven't our boys to work for, we'll work for our country' still,
Be glad that your men are fighters – for the shame that surely hurts
Is to have a coward man-child, who hides behind women's skirts.
Just clench your teeth when you read the lists or the wounded and
    the dead,
And if the names that you love are there – be proud and hold up
    your heed,
Don't cry! For they've climbed the pathway that heroes and martyrs
    trod:
They sleep in the Rest of Heaven! They stand in the Glory of God!

Notes

1 Samuel Steele was born on 15 February 1837 and died on 5 November 1912. Mary Maria, née Bennion, was born 16 June 1835 and died on 23 December 1903. They married on 30 January 1861 and produced seven children: Samuel, Francis Mary (Laura), Nelly, James Reginald (died in infancy), Herbert Harry (died in infancy), Ethel May, Beatrice Ada, and John Rawdon.

2 *Evening Telegram*, 11 September 1886. The firm is described as a dry goods and grocery store located in "Manchester House," a premises that formerly belonged to Messrs L. O'Brien & Co., connected at one time with the prosecution of the seal and cod fishery. The paper noted in particular that Steele had brought in a large stock of elegant cloaks of the latest London and Paris styles and colours.

3 *Evening Telegram*, 5 September 1889.

4 O'Neill, *A Seaport Legacy*, 2:867–8.

5 The children were Owen William (28 April 1887), James Robert (19 July 1888), Herbert Harris (4 September 1889), Mary Blanche (18 January 1891), Beatrice Grace (25 June 1892), Sarah Frances (14 July 1893), Samuel Richard (5 March 1895), Ethel May (20 July 1886), Victor John (31 December 1897) and Ella Gertrude (28 July 1899).

6 *Who's Who In and From Newfoundland,* 1927, 257–8.
7 Nicholson, *More Fighting Newfoundlanders,* 115.

INTRODUCTION

1 Hynes, *The Soldiers' Tale,* xi.
2 Ibid., xv, xvi.
3 Ibid., xiv, 2, 4–5.
4 These include, of course, the Owen Steele diary and letters, the James
   Steele diary and post-war correspondence, the letters of James H. Carter,
   the Anthony Stacey memoirs, the Lewis Bartlett papers, the Rupert
   Bartlett letters, the letters of Lester Barbour, the H.G.R. Mews correspondence, and transcripts of CBC interviews with World War One veterans
   Abe Mullett, Lt. Col Hadow, Lt. Col O'Driscoll, Howard Morry, and
   Major A. Raley. With the exception of the Bartlett Papers (which the
   author has had an opportunity to examine) and the CBC transcripts, this
   material is located in original or xerox form in the Archives of the Center
   for Newfoundland Studies at Memorial University of Newfoundland. At
   the time of writing, the papers of the former regimental chaplain, Lt. Col.
   T.F. Nangle, were being catalogued by the Archives. The letters of two
   Newfoundland women who served with the Volunteer Aid Detachment
   (VAD), Frances Cluett and Sybil Johnson, also form part of the collection.
5 Alonzo John Gallishaw (1890-?), writer and educator, was born in St.
   John's and educated at Harvard University. When war broke out in 1914
   Gallishaw crossed the border and signed up with the Canadian Army. He
   later requested and received a discharge after which he returned to Newfoundland and enlisted with the Newfoundland Regiment. He worked for
   a while in London looking after the regiment's records for the War
   Office. When the regiment headed for Gallipoli in 1915, he managed to
   stow himself on board the train and then the transport bound for Malta.
   He persuaded the regimental adjutant to assign him to B Company where
   he saw service on the peninsula. He was wounded during the evacuation
   and subsequently demobilized. He returned to Harvard as a lecturer and,
   when the United States entered the war in 1917, served with American
   forces in France. He commanded a battalion briefly before taking up an
   appointment with Army Intelligence and was later a liaison officer with
   the British Army. He published five books, and also wrote plays, radio
   and television scripts, and worked for several Hollywood studios as a
   script consultant. Smallwood, *Encyclopedia of Newfoundland and
   Labrador,* 2: 463–4.
6 Nicholson, *The Fighting Newfoundlander,* 296.
7 O'Brien, "The Newfoundland Patriotic Association."

8   Lind, *The Letters of Mayo Lind*.
9   Hynes, *Soldier's Tale*, 11.
10 For details see Penney, "The Church Lads Brigade and the Catholic Cadet Corps."
11 Nicholson, *The Fighting Newfoundlander*, 97.
12 Ibid., 92, 96.
13 O'Brien, "The Newfoundland Patriotic Association," 29.
14 Ibid., 73.
15 Ibid., 169–71.
16 *The Dissenting Church of Christ in St. John's, 1775–1975*, 114, 185.
17 National Archives of Canada (NAC) 150/Accession 1992-92/171, Box 5, File 0-89. Attestation Papers, Department of Militia, 25 March 1919; Extract from Nominal Roll. Embarked St. John's per ss "Florizel" October 4, 1914.
18 Hynes, *Soldier's Tale*, xiii.
19 Ibid., 2.
20 Nicholson, *The Fighting Newfoundlander*, 276.
21 Provincial Archives of Newfoundland and Labrador (PANL), *The Veteran*, April 1921 to January 1924. Bound volumes. Vol. 1, no. 1, D.L. George "Introduction." For a more recent and through analysis of enlistments and casualties in Newfoundland's armed forces see Sharpe, "The Race of Honour," 27-55.
22 Nicholson, *The Fighting Newfoundlander*, 423; Noel, *Politics in Newfoundland*, 122.
23 Noel, *Politics*, 24–5, 41–6, 58, 62–3.
24 Ibid., 4.
25 Ryan, "Newfoundland Spring Sealing Disasters to 1914," 40–4.
26 Noel, *Politics*, 8-9; Sharpe, "The Race of Honour," 38.
27 Noel, *Politics*, 21; McDonald, "To Each His Own," 38. For detail on the role of sectarian groups in the war effort see Churchill, "No Surrender until Prussian militarism is crushed."
28 *The Veteran*. April 1921 to January 1924. Bound volumes. Vol. 1, no. 2 (April, 1921), 11.
29 Brown, "Citizen Soldiers: Political Pawns." All the British North American colonies at the time had volunteer militia units, partly in response to a revival of militarism in the late 1850s but also to provide local forces to support the British regulars if the need arose. The volunteer structure in Canada, Nova Scotia, and New Brunswick laid the foundation for the Canadian militia following Confederation. See Facey-Crowther "Militiamen and Volunteers."
30 *Calypso* was one of the last of the Royal Navy's sailing corvettes. It was sent to Newfoundland in 1889 to be used as a training ship and depot

for the Royal Newfoundland Naval Reserve. It was renamed HMS *Briton* in 1916. See O'Neill, *A Seaport Legacy*, 975; Sharpe, "Race of Honour," 42–3.

31   Sharpe, "The Race of Honour," 38.

32   Lucas, *The Empire at War*, 2: 297–8.

33   McDonald, *Coaker*, 48, 60.

34   Noel, *Politics*, 130.

35   *The Daily News*, 5 August 1914.

36   Sir Walter Edward Davidson (1859–1923), governor of Newfoundland (1913–1917). Educated in Cambridge, England. In 1880, he took up an appointment with the Colonial Service in what was then called Ceylon (Sri Lanka). He remained there until 1901 when he moved to Transvaal, where he held a number of positions. In 1912, he was appointed governor of the Seychelles Islands and, in 1913, governor of Newfoundland. He left Newfoundland in 1917. His last appointment, before retirement in 1923, was as governor of New South Wales. He died in 1923. Smallwood, *Encyclopedia of Newfoundland and Labrador*, 1: 595.

37   McDonald, *Coaker*, 48.

38   Nicholson, *The Fighting Newfoundlander*, 103-4. The war effort extended well beyond St. John's and engaged hundreds of women across Newfoundland and Labrador in raising money and providing comforts for those serving with the forces overseas. See Warren, "The Patriotic Association of the Women of Newfoundland."

39   O'Brien, "The Newfoundland Patriotic Association."

40   Gilbert, "Newfoundland Regiment." Archives, Center for Newfoundland Studies (CNS), Memorial University of Newfoundland. Newfoundland Patriotic Association. Letters and Statistics 1917, 25; Atwater, "Echoes and Voices Summoned from a Half-Hour in Hell," 200–1.

41   PANL. W.F.R., "The Blue Puttees," 13–15; Nicholson, *The Fighting Newfoundlander*, 107–10.

42   Nicholson, *The Fighting Newfoundlander*, 106.

43   Under the Military Service Act of April 1918 an additional 1,573 were conscripted into the Regiment. Sharpe, "The Race of Honour," 27–9.

44   Nicholson, *The Fighting Newfoundlander*, 117.

45   *The Daily News*, 4 October 1914.

46   Archives, CNS MUN, MF 147. Owen W. Steele, "Diary of the Newfoundland Regiment," 4, 6–7, 64.

47   For details on the problems faced by the regiment in dealing with battle attrition see Parsons, "Morale and Cohesion in the Royal Newfoundland Regiment 1914–18," chapter 3.

48   O'Brien, "The Newfoundland Patriotic Association."

49   Ibid., 336-7.

50   Ibid., 336.

CHAPTER ONE

1 The *Florizel* had a tragic history. Six months earlier it had been involved in the great Newfoundland sealing disaster when seventy-eight men froze to death on the ice floes. Many of the first 500 it conveyed across the Atlantic were killed or wounded at Beaumont Hamel. In February 1918 it ran aground during a storm off the Newfoundland coast with a loss of ninety-four lives. Dancocks, *Welcome to Flanders Fields*, 79, 79n.
2 Pritchett, *A History of the Church Lads' Brigade in Newfoundland. 100th Anniversary*, 1–6. Note in particular the emphasis on infantry training in the 1910 *Guide for Officers, Warrant Officers, Non-Commissioned Officers, and Privates of the Church Lads' Brigades Infantry Training*. See also Nicholson, *The Fighting Newfoundlander*, 92–6.
3 This a point developed by Andrew Parsons in his "Morale and Cohesion in the Royal Newfoundland Regiment 1914–1918."
4 This event is recounted in Nicholson, *The Fighting Newfoundlander*, 115–6.
5 Ryan, *The Ice Hunters*.
6 Cramm in his *The First Five Hundred*, 25 describes the embarkation of the First Contingent for England. See also Gallishaw, *Trenching at Gallipoli*, 5.
7 The thirty-two grey-painted ships of the Canadian convoy left Quebec City on 3 October, carrying over 30,000 men. They were joined off Newfoundland by the battleship HMS *Glory* and the cruiser *Lancaster*. On 10 October the convoy was joined by another battleship, the *Majestic* and the Royal Navy's newest battle cruiser *Princess Royal*. See Dancocks, *Flanders Fields*, 74–9.
8 The *Florizel* took position as the last transport in the port column headed by S.S. *Megantic*. See Nicholson, *The Fighting Newfoundlander*, 117.
9 The rank markings were three stripes, a crown and two crosses flags.
10 The Canadians were transporting 7679 horses with their contingent. Presumably these were casualties from the rough crossing. See Dancocks, *Flanders Fields*, 74.
11 Details of the concert, with a mention that "its success owed much to the organizing ability of Colour-Sergeant (Company Sergeant-Major) Owen Steele," are found in Nicholson, *The Fighting Newfoundlander*, 120.
12 George Thomas Carty, granted commission, Captain 21 September 1914; BMEF 20 August 1915; evacuated Suvla, sick, 4 December 1915; admitted to hospital, Malta, 8 December 1915; invalided to England, 16 April 1916; embarked for Newfoundland, 9 June 1915; appointed Transport Officer, 8 July 1916; Major, 1 October 1916; appointed officer commanding depot, St. John's, 29 October 1917; relinquished command, 22 July 1918; Lieutenant-Colonel, 1 January 1919; appointed staff officer, acting DOC, 28 January 1918; relinquished appointment as acting DOC, 9

September 1919; retired, 9 September 1919. Cramm, *The First Five Hundred*, 124. Photograph, 125.

13 Private W.D. Stenlake was a Methodist student from the Twillingate area. He eventually served as the regiment's Protestant padre in France. He won the admiration of the soldiers he served with. Mayo Lind described Stenlake as "one of the best that ever lived." *The Letters of Mayo Lind*, 28–9, 50. Nicholson, *The Fighting Newfoundlander*, 119, 149, 177, 528.

14 Dr Arthur William Wakefield served for many years with the Grenfell Mission in St. Anthony. He was founder of the senior unit of the Legion of Frontiersmen in St Anthony. He was commissioned Lieutenant in the regiment on 21 September 1914; struck off strength, Aldershot 2 October 1915; transferred to Royal Army Medical Corps. See Cramm, *First Five Hundred*, 131; Nicholson, *The Fighting Newfoundlander,* 97–8. There is a reference to him in *The Letters of Mayo Lind*, 70, where he explains the use of the gas helmet to members of the regiment.

15 The 27,000 ton *Princess Royal* was one of the Royal Navy's newest and fastest battle cruisers. See Dancocks *Flanders Fields*, 77, 79.

16 Nicholson, *The Fighting Newfoundlander*, 120.

17 This may refer to the checking of the British advance into Flanders in the early part of the month. See John Terraine, *The First World War 1914–18*, 41.

18 Presumably the fall of Antwerp and the withdrawal of the Belgian Army down the coast where it would eventually link up with Entente forces. See Terraine, *The First World War*, 41

19 William Hodgson Franklin. Franklin was one of the senior officers of the Church Lads' Brigade before the war. Commissioned Captain 21 September 1914. He served as Camp Commandant while the regiment was forming at Pleasantville, Newfoundland. Embarked for u.k. 2 October 1914. Franklin commanded the first contingent that left St John's in October 1914, although he left beforehand on the *Carthaginian* to make arrangements for clothing and equipment for the Newfoundlanders. Attached to the First Battalion Suffolk Regiment, 18 November 1914; to the 6th Battalion, Royal Warwick Regiment, 20 March 1915; to the BEF 20 March 1915. Major 25 March 1915; Lieutenant-Colonel, 15 April 1916; in command of 6th Royal Warwick Regiment; wounded Beaumont Hamel, 1 July 1916; awarded DSO July 1, 1916; invalided to England, 6 July 1916; mentioned in despatches, 5 January 1917; retired, u.k., with honorary rank of Colonel, 29 November 1919. Cramm, *The First Five Hundred*, 124; see Nicholson, *The Fighting Newfoundlander*, 93, 106, 114, 119, 154.

20 Henry LeMessurier. Civil servant, politician, author, composer of song "The Ryans and the Pittmans." Sometime editor of the *Evening Herald*; deputy minister of Customs Department. He was also a prolific writer

and lecturer on historical topics. Smallwood, *Encylopedia of Newfound-land and Labrador,* 279.

21 The *Florizel* reached Plymouth Sound on the 14th but the First Five Hundred did not disembark until 20 October. It had been originally planned for the convoy to go to Southampton, but reports of German submarines in the English Channel changed the destination at the last minute to Plymouth and Devonport. Nicholson, *The Fighting Newfoundlander,* 121.

22 Nicholson, *The Fighting Newfoundlander* 121–2. ; Gallishaw. *Trenching in Gallipoli,* 5.

23 This is Nelly Letten, S.O. Steele's sister, who lived in Gravesend, Kent.

24 The soldiers complained that the tents were without wooden floors; these were promised for delivery in a few days but in the end it turned out to be several weeks. Nicholson, *The Fighting Newfoundlander,* 122.

25 Laura Bottomley, S.O. Steele's sister, whose husband Frank Bottomley was manager of the MacIntosh Toffee Company in Yorkshire.

26 Frederick Sleigh, 1st Earl Roberts of Kandahar, Pretoria, and Waterford (1832-1914). British military commander who distinguished himself in the second Afghan and Boer wars. Following his successes in the Boer War, he returned to Britain and, a hero, was created an earl and appointed commander in chief of the British army. He retired in 1904. *Microsoft Encarta Encyclopedia* 2001.

27 Possibly Robert Templeton Smith. Reg. No. 366. Enlisted 4 September 1914; second lieutenant, Seaforth Highlanders, and struck off strength 24 October 1915. See Cramm, *The First Five Hundred,* 286.

28 Charles E.A. Jeffery (1881–1964) had been assistant master of Feild College until 1913. On the outbreak of war he joined the Canadian Expeditionary Force. In 1917 he was appointed to the Canadian High Commisioner's Office in London. After the war he served briefly as principal of the Soldier's Re-Establishment School and in 1921 rejoined the teaching staff of Bishop Feild College. He served as editor of the *Evening Telegram* from 1923 until 1959. For his contribution to the civic life of St John's he was awarded the MBE by King George VI in 1936. Smallwood and Pitt, *Encyclopedia of Newfoundland and Labrador,* 3:103–4.

29 Private Reginald Grant Patterson. Enlisted 7 September 1914; Lance Corporal 30 January 1915; Second Lieutenant 22 April 1915; Lieutenant 16 October 1915; BMEF 14 November 1915; BEF 14 March 1916. Wounded, Beaumont Hamel, 1 July 1916; invalided to England 4 July 1916; attached to depot, Ayr, 10 August 1916. Returned to BEF, awarded Military Cross, 26 September 1917; awarded Bar to Military Cross, 20 November 1917; wounded Ledegham 14 October 1918; invalided to England, 18 October 1918; attached to depot, Winchester, 14 February 1918; embarked for Newfoundland, 15 February 1919; retired, St John's,

19 March 1919. Cramm, *The First Five Hundred*, 264; Nicholson, *The Fighting Newfoundlander*, 386.

30 This refers to the colours of the so-called "native" flag of Newfoundland with its vertical panels of pink, white, and green. The flag originated in the 1800s out of divisions between citizens whose family were long connected with St John's and Newfoundland and recently arrived immigrants from Ireland. It never received official status but is still flown by some residents of the province on public occasions. Smallwood and Pitt, *Encyclopedia of Newfoundland and Labrador*, 2: 196–7.

31 Probably the 5th Royal Highlanders of Canada, one of Montreal's militia regiments. See Dancocks, *Flanders Fields*, 38.

32 Captain John Wesley March. Commissioned Captain Newfoundland Contingent, 21 September 1914; served with BMEF; awarded Military Cross, 11 December 1916; evacuated and invalided to England, 21 January 1917; attached to depot, Ayr, 14 March 1917; awarded French Croix de Guerre, 1 May 1917; Major, 1 July 1917; embarked for Newfoundland, 22 May 1919; retired, 5 July 1919. See Cramm, *The First Five Hundred*, 127.

33 Bertram Butler. Reg. No. 146. Enlisted 2 September 1914; Orderly Room Sergeant, 3 September 1914; Lieutenant, 4 October 1914; served with . BMEF, 20 August 1915; Captain, 10 January 1916; wounded three times; transferred to England, 27 November 1917; awarded Military Cross, Bar to Military Cross, Distinguished Service Order; mentioned in Sir Douglas Haig's despatches; returned to Newfoundland 2 September 1918; Major, 15 September 1919; seconded for duty with Civil Re-Establishment Committee 14 January 1920. See Cramm, *The First Five Hundred*, 12.

Butler returned to service during World War II and was stationed in Corner Brook, Newfoundland, with the Home Defence Battery. In the post-war years he served as Honorary Lieutenant Colonel of the Royal Newfoundland Regiment. Smallwood and Pitt, *Encyclopedia of Newfoundland and Labrador*. 1: 301, 446.

34 Lieutenant-General Edwin Alderson, a veteran of the British Army who had commanded the Canadians in the Boer War. Alderson was replaced by Lieutenant-General Sir Julian Byng on 29 May 1916. Dancocks, *Flanders Fields*, 83, 339.

35 Field Marshal Frederick, Earl Roberts (1832–1914).

36 Robert Holland Tait was son of Dr James Tait. Granted Commission Lieutenant 24 September 1914; BMEF 8 August 1915; admitted to hospital 8 September 1915, discharged 4 October 1915; embarked for Gallipoli, 19 October 1915; Captain, 10 January 1916; admitted to hospital, Cairo, 28 February 1916; discharged to duty, 30 April 1916; BEF, 16 May 1916; evacuated to hospital, 20 May 1916; discharged to duty, 19 October 1916; appointed Adjutant, 23 May 1917; evacuated to hospital,

17 July 1917; rejoined the Battalion, 24 August 1917; awarded Military Cross, 9 October 1917; wounded, Neuve Eglise, 12 April 1918; invalided to England, 15 April 1918; embarked for Newfoundland on special duty, 22 May 1918; appointed acting Major, 1 June 1919, while commanding discharge depot, St John's; retired, 6 October 1919. Cramm, *The First Five Hundred*, 131. See also Lind, *The Letters*, 28. Earlier, when the Regiment was at Fort George, Invernessshire, Scotland, Lieutenant Tait organized a recreation room for the Regiment. In exchange for 3d. (6 cents) a month the soldiers were supplied with writing utensils, games, newspapers, and magazines.

37 The Newfoundland contingent, the Fourth Canadian Infantry Brigade, and two Canadian calvary units were located approximately eight miles north-west of Stonehenge. Nicholson, *The Fighting Newfoundlander*, 121.

38 Conn Alexander. Granted Commission, Captain, 21 September 1914; BMEF, 20 August 1915; BEF, 14 March 1916; relinquished Commission, 20 June 1916. Cramm, *The First Five Hundred*, 123.

39 Canadian headquarters was established at Bustard Camp, about three miles north of Stonehenge. Nicholson, *The Fighting Newfoundlander*, 121.

40 This may be William Martin, who came out with his brother Frank from Devon, England, to take charge of J.H. Martin & Co. before the Great Fire. See O'Neill, *A Seaport Legacy*, 881.

41 Lord Brassey, former first lord of the admiralty, was invited to join the Newfoundland War Contingent Association because of his interest in Newfoundland. Nicholson, *The Fighting Newfoundlander*, 127.

42 A nautical term used to describe a method of fastening large ropes by a half hitch. Combined with keg, as in clinch keg, it refers to a wooden float attached at the point where the anchor-chain or boat mooring is clinched to a cable that leads to a buoy. Dictionary of Newfoundland web site, www.heritage.nf.ca/dictionary/d8ction.html. Not clear why Steele used it here.

43 Laurie Graham Baine. Enlisted, 8 September 1914; Lance Corporal, 12 March 1915; 2nd Lieutenant, 6 April 1915; BMEF, 20 August 1915; Lieutenant, 15 October 1915; evacuated Suvla, sick, 23 November 1915; invalided to England, 30 November 1915; attached to Royal Flying Corps, 25 September 1916; graded Flying Officer, 30 March 1917; BEF with the Royal Flying Corps, 2 June 1917; Captain, 1 April 1918; evacuated to hospital, 24 October 1918; transferred to England, 5 November 1918; rejoined 2nd Battalion, Winchester, 4 March 1919; embarked for Newfoundland, 2 May 1919; retired, 21 May 1919. Cramm, *The First Five Hundred*, 135.

44 These are Frank and Laura Bottomley and their children.

45 A.B. Young was employed by and later became one of the partners in

David Metheun and Sons Pottery firm in Kirkcaldy, Scotland, and was a
family friend. Presumably the firm supplied S.O. Steele with china items.
These facts are gleaned from several letters from Young to Jim Steele
while the latter was overseas, which are now in the possession of his son,
James Steele.

CHAPTER TWO

1 From 13 Dec. 1914 until 19 Aug. 1915, there are no diary entries. We
have only the correspondence for information about this period.

2 Nicholson, *The Fighting Newfoundlander*, 128–32.

3 Lieutenant-Colonel Richard de H. Burton took over command of the regiment in Bustard Camp on Salisbury Plain in November 1914. He was a
British regular officer, fifty-three years old, and brought out of retirement
on the outbreak of war. He had served in South Africa with the 77th
Foot, the Duke of Cambridge's Own (Middlesex Regiment). He was a
soldier of the old school, a firm believer in time-tested methods of getting
men in shape. This meant long route marches to toughen feet interspersed with sessions of drill and musketry instruction. See Nicholson,
The Fighting Newfoundlander, 125.

4 Francis Herbert Knight. Enlisted 8 September 1914; Lance Corporal, 15
April 1915; 2nd Lieutenant, 22 April 1915; BMEF, 20 August 1915; Lieu-
tenant, 1 Jan 1916; BEF, 14 March 1916; wounded, Beaumont Hamel, 1
July 1916; invalided to England, 11 July 1916; returned to Newfound-
land, 20 July 1916; retired, 22 December 1916. Cramm, *The First Five
Hundred*, 220.

5 Michael Vail. Enlisted 5 September 1914; Lance Corporal, 15 April
1915; Corporal, 2 July 1915; BMEF, 20 August 1915; evacuated from
Suvla, 23 October 1915; invalided to England, 12 April 1916; repatriated
to Newfoundland, 30 June 1916; discharged, St John's, medically unfit,
13 October 1916. Cramm, *The First Five Hundred*, 302.

6 James Mifflin. Enlisted 3 September 1914; Lance Corporal, 14 June
1915; BMEF, 20 August 1915; evacuated Suvla, 17 October 1915;
invalided to England, 1 November 1915; Corporal, 20 April 1915;
attached to depot, Ayr, 28 April 1916; Acting Sergeant, 11 August 1916;
2nd Lieutenant, 1 November 1916; BEF, 23 April 1917; Lieutenant, 18
March 1918; wounded, Haslar Camp, Belgium, 27 March 1918; awarded Military Cross, 29 March 1918; invalided to England, 3 April 1918;
discharged from Hospital, 22 June 1918; embarked for Newfoundland,
furlough, 21 July 1918; returned to United Kingdom, 19 October 1918;
returned to BEF, 19 November 1918; rejoined Battalion, 26 November
1918; embarked for Newfoundland, 22 May 1919; retired and placed on
Reserve of Officers, 20 June 1919. Cramm, *The First Five Hundred*, 244.

7 This may be either Frederick LeGrow or Frederick Pratt LeGrow.
   Cramm, *The First Five Hundred*, 224–5.

8 George Paver. Enlisted 4 September 1914; Regimental Sergeant Major 21
   September 1915; BMEF, 20 August 1915; admitted 5th Canadian Station-
   ary Hospital, Cairo, 12 September 1915; discharged from Hospital, 17
   October 1915; rejoined Battalion, Suez, 31 January 1916; BEF, 14 March
   1916; evacuated to hospital, April 1916; classified "PB" (Permanent
   Base) at Base Depot, Rouen, 4 April 1916; transferred to England, 27
   April 1916; attached to depot, Ayr, 14 June 1916; embarked for New-
   foundland, 20 July 1916; discharged, St John's, medically unfit, 11
   December 1916. Cramm, *The First Five Hundred*, 264.

9 Malcolm Godden. Enlisted 2 September 1914; Corporal, 24 September
   1914; Sergeant, 10 October 1914; Company Sergeant Major, 30 July
   1915; BMEF, 20 August 1915; BEF, 14 March 1916; evacuated to hospital, 23 August 1916; invalided to England, 4 September 1916; attached
   to depot, Ayr, 6 November 1916; 2nd Lieutenant, 1 November 1916;
   Lieutenant, 1 May 1918; embarked for Newfoundland on furlough, 21
   July 1918; returned to United Kingdom, 19 October 1918; evacuated to
   hospital, 28 November 1918; attached to 2nd Battalion, Winchester, 28
   April 1919; embarked for Newfoundland, 5 September 1919; retired 14
   October 1919. Cramm, *The First Five Hundred*, 184.

10 Walter Frederick Rendell. Granted Commission, Captain, 21 September
   1914; BMEF, 20 August 1915; wounded, Suvla Bay, 20 September 1915;
   admitted hospital, Malta, 26 September 1915; invalided to England, 8
   October 1915; attached to depot, Ayr, 8 February 1916; Major, 25 April
   1916; BEF, 5 September 1916; wounded, Gueudecourt, 12 October 1916;
   invalided to England, 18 October 1916; attached to depot, 20 February
   1917; embarked for Newfoundland on special duty, 21 August 1917;
   Appointed Chief Staff Officer, Department of Militia, 29 October 1917;
   Lieutenant-Colonel, 16 May 1918; awarded CBE 12 December 1919;
   remained on strength of regiment after the war. Cramm, *The First Five
   Hundred*, 128.

11 The Second Contingent actually sailed to Britain on the *Dominion*. See
   Nicholson, *The Fighting Newfoundlander*, 206.

12 Nicholson, *The Fighting Newfoundlander*, 136–7.

13 Arthur Raley. Before enlistment Arthur Raley was games master at Bishop
   Feild College, a Church of England school, and was an officer in the
   Church Lads' Brigade. Granted commission, Lieutenant, 24 September
   1914; Captain, 28 July 1915; BMEF, 20 August 1915; Appointed Adjutant,
   29 September 1915; admitted hospital, Malta, 13 December 1915; rejoined
   Battalion, Suez, 31 January 1916; BEF, 14 March 1916; awarded Military
   Cross, 1 January 1917; wounded, Monchy, 14 April 1917; invalided to
   England, 19 April 1917; attached to depot, Ayr, 23 May 1917; returned to

BEF 5 January 1918; rejoined Battalion, 26 January 1918; appointed Adjutant, 24 June 1918; Temporary Major, 5 January 1919; transferred to England for Officers' Course, 11 January 1919; Rejoined 1st Battalion, Winchester, 23 April 1919; appointed Second in Command of 1st Battalion, Winchester, 20 May 1919; embarked for Newfoundland, 22 May 1919; decorated with Croix de Guerre with Silver Star, 19 June 1919; retired 29 July 1919. Cramm, *The First Five Hundred*, 128; Nicholson, *The Fighting Newfoundlander*, 112, 171 and Appendix B.

14 Norman Harvey Alderdice. Granted commission Lieutenant 24 September 1914; relinquished commission 8 October 1915 and joined a unit of the Imperial Army. Cramm, *The First Five Hundred*, 123.

15 Charles Wighton. Granted commission Lieutenant, 21 September 1914; BMEF, 20 August 1915; Captain, 17 October 1915; killed in action, Suvla Bay, 25 November 1915. Cramm, *The First Five Hundred*, 131, photograph, 130. Wighton had received his captaincy in mid-October, 1915. Nicholson, *The Fighting Newfoundlander* 181.

16 Dame Gladys Cooper (1888–1971). British actress and theatre manager. Born in 1888 in Lewisham, England. Her long career as an actress began at the young age of sixteen and continued until the mid 1960s. She performed on both the London and New York stage to great acclaim and then in film where she received three nominations for Academy Awards for best actress. For a brief period, early in her career, she also managed the London Playhouse Theatre (1927–33). She wrote two volumes of memoirs, *Gladys Cooper* (1931) and *Without Veils* (1953). She died in 1971. *Microsoft Encarta Encyclopedia 2001*

17 Seymour Hicks (1871–1949) performed in the first stage review, *Under the Clock*, with Charles Brookfield. The 1920s revues, and the one attended by Steele was perhaps one of these, were spectacular entertainments. Hicks was later knighted and as Sir Seymour Hicks played his second version (the first had been a 1913 silent movie) of Charles Dicken's *Scrooge* in 1935. "Revue," Murray, *The Hutchinson Dictionary of the Arts*.

18 Sir John Harvey-Martin was born John Martin Harvey on 22 June 1863 at Wivenhoe, Essex, England. Son of a yacht builder, Harvey originally intended to undertake a career in naval architecture but instead studied theatre with the actor John Ryder. He made his first appearance on the London stage in 1881. The following year he joined the Lyceum Theatre company of Sir Henry Irving, remaining there for fourteen years and traveling frequently to the United States. His greatest success was in *The Only Way* (1899). He subsequently performed in a number of Shakespearean productions. One of his finest performances was in the London production of *Oedipus Rex* in 1912. He was knighted in 1921 at which point he added the hyphen to his name. He died in 1944. Murray, *The*

Hutchinson Dictionary of the Arts; see also "Sir John Martin-Harvey" on web.ukonline.co.uk/thornbury/jmh.html.

19 George Langmead Jr. Enlisted 2 September 1914; Sergeant, 21 September 1914; Company Quartermaster Sergeant, 31 July 1915; BMEF, 20 August, 1915; BEF, 14 March 1916; wounded, Beaumont Hamel, 1 July 1916; commissioned 2nd Lieutenant, 1 July 1916; invalided to England, 4 July 1916; BEF, 12 March 1917; Lieutenant, 1 November 1917; wounded, Mesnieres, 2 December 1917; died of wounds, 8 December 1917. Cramm, *The First Five Hundred*, 223.

20 Lamont Paterson. Commissioned Captain, 21 September 1914; Major, 3 June 1916; appointed Deputy Director of Medical Services, 14 November 1916; Lieutenant-Colonel, 7 December 1918; OBE, 3 June 1919; relinquished appointment DDMS, 9 September 1919; retired, 9 September 1919. Cramm, *The First Five Hundred*, 128.

21 Alexander Montgomerie. Commissioned Captain, September 1914. Alexander was Conducting Officer for the first draft of the regiment that left for Britain in October, after which he returned to his duties as Adjutant of the Regimental Depot at St John's. Montgomerie and Lieutenant Eric S. Ayre were the two service representatives on the Reserve Force Committee. This committee, under the direction of the Newfoundland Patriotic Association, was responsible for enrolling and training drafts for the regiment. Montgomerie was also Conducting Officer for several of the drafts sent to England in 1915, including G Company which he commanded temporarily after its arrival in Scotland. He was promoted to the rank of Major and took on the post of Officer Commanding the Depot at St John's in June 1916. In August 1917, as District Officer Commanding, he took responsibility for administering all matters relating to the military establishment in Newfoundland. Nicholson, *The Fighting Newfoundlander*, 108, 137, 197, 206, 212, 216, 368–9, 403.

22 Augustus O'Brien. Commissioned Captain, 21 September 1914; BMEF, 20 August 1915; evacuated Suvla, sick, 21 December 1915; admitted hospital, Malta, 18 January 1916; rejoined Battalion, Suez, 7 March 1916; wounded, Gueudecourt, 12 October 1916; died of wounds, 18 October 1916; mentioned in despatches, 5 January 1917. Cramm, *The First Five Hundred*, 127.

23 Charles Robert Ayre. Commissioned Lieutenant, 24 September 1914; Captain, 28 July 1915; returned to Newfoundland and attached for duty to depot, 8 May 1916; retired, St John's, medically unfit, 7 June 1917. Cramm, *The First Five Hundred*, 123.

24 Joe Nunns was granted Commission, Lieutenant, 24 September 1914; BMEF, 20 August 1915; Captain, 10 January 1916; BEF, 14 March 1916; wounded Beaumont Hamel, 1 July 1916; invalided to England, 3 July 1916; attached to depot at Ayr, 13 September 1916; returned to BEF, 9 January 1917; evacuated to hospital, 13 March 1917; discharged to duty,

11 May 1917; awarded Military Cross, 1 October 1917; returned to United Kingdom and substituted for rest after twenty-four months continuous service in the field; attached to Depot Winchester, 5 April 1918; embarked for Newfoundland on special duty, 13 May 1918; embarked for United Kingdom, 12 October 1918; attached to Depot Winchester, 22 October 1918; returned to Newfoundland, 12 December 1918; retired, 1 February 1919. Cramm, *The First Five Hundred* 127, photograph 126. Captain Herbert Cramm, along with Nunns and four other men on the Newfoundland Regiment were the last of the Allied Troops to leave the Gallipoli peninsula. See Cramm, *First Five Hundred*, 47.

25 William Hoyes Grant. Enlisted 11 September 1914; Lance Corporal, 21 May 1915; 2nd Lieutenant, 16 October 1915; BEF, 23 March 1916; killed in action, in the line near Beaumont Hamel, 16 July 1916. Cramm, *The First Five Hundred*, 188.

26 Robin Stick. Enlisted, 2 September 1914; Lance Corporal, 13 November 1914; Corporal, 7 April 1915; Sergeant, 14 June 1915; Second Lieutenant, 29 July 1915; BMEF, 20 August 1916; BEF, 14 March 1916; wounded, Beaumont Hamel, 1 July 1916; invalided to England, 4 July 1916; attached to depot, Ayr, 17 October 1916; returned to BEF, 12 March 1917; Captain, 6 August 1917; wounded, Marcoing, 20 November 1917; discharged to duty, 30 November 1917; embarked for Newfoundland, furlough, 24 July 1918; returned to United Kingdom, 19 October 1918; returned to BEF, 4 November 1918; embarked for Newfoundland, 22 May 1919; retired and placed on reserve of officers, 6 June 1919. Cramm, *The First Five Hundred*, 293.

27 John Edward Joseph Fox. Enlisted 4 September 1914; Lance Corporal, 21 September 1914; Corporal, 13 November 1914; Sergeant, 24 February 1915; 2nd Lieutenant, 29 March 1915; BMEF, 20 August 1915; Lieutenant, 15 October 1915; evacuated, Suvla, sick, 3 November 1915; rejoined Battalion, Suez, 26 January 1916; BEF, 14 March 1916; evacuated to England, sick, 27 April 1916; Captain, 16 November 1916; appointed Demobilization Officer, London, 11 November 1918; returned to Newfoundland, 22 May 1919; relinquished appointment, 2 June 1919; retired, 23 June 1919. Cramm, *The First Five Hundred*, 180.

28 Edward Patricks Morris (1859–1935). Lawyer and prime minister of Newfoundland. Born in St John's and educated there at St Bonaventure's College and at the University of Ottawa. He was prime minister of Newfoundland from 3 March 1909 to 31 December 1917. He resigned his office while attending the Imperial War Cabinet in London. In 1918 he was created the first Baron Morris of St. John's and Waterford – the only native Newfoundlander to be elevated to the British peerage. He remained in London thereafter where he died on 24 October 1935. Smallwood, *Encyclopedia of Newfoundland and Labrador*, 3: 83–5.

29 John Henry Stanley Green. Enlisted, 2 September 1914; BMEF, 20 August 1915; evacuated to England, sick, 29 October 1915; 2nd Lieutenant, 12 July 1916; attached to 57th Squadron Royal Flying Corps as Flying Officer, 12 March 1917; BEF, 21 March 1917; killed in action, 7 July 1917. Cramm, *The First Five Hundred*, 188.

30 Prominent St. John's merchant, member of the Newfoundland Patriotic Committee, Bowring presented the Newfoundland Regiment with a set of band instruments. See Nicholson, *The Fighting Newfoundlander*, 104, 217, 220, 224, 336, 515–16.

31 Michael Francis Summers. Appointed Quartermaster, 21 September 1914; BMEF, 20 August 1915; Captain, 23 November 1915; BEF, 14 March 1916; wounded, Beaumont Hamel, 1 July 1916; died of wounds, 16 July 1916. Cramm, *The First Five Hundred*, 131.

32 Nicholson, *The Fighting Newfoundlander*, 151.

33 Lind, *The Letters*, 35ff; Nicholson, *The Fighting Newfoundlander*, 150–1.

34 James Allan Ledingham. Granted commission, Lieutenant, 2 September 1914; Captain, 28 July 1915; BEF, 13 March 1916; wounded, Beaumont Hamel, 1 July 1916; invalided to England, 3 July 1916; attached to depot, Ayr, 14 September 1916; returned to BEF, 21 December 1916; appointed Adjutant, 2 December 1916; rejoined Battalion, 15 May 1917; killed in action, Broembeck, 9 October 1917. Cramm, *The First Five Hundred*, 127.

35 Captain Wilfred J. Pippy of Springdale, St. John's, Newfoundland. Enlisted 26 January 1915 and commissioned lieutenant; served with the regiment until 6 June 1916 when he was hospitalized for nephritis; released from hospital on 4 March 1917 but deemed unfit for active service and repatriated 4 May 1917; received the temporary rank of Captain on 1 October 1916, made substantive on 2 July 1918; rejoined regiment in the field, 22 November 1918; returned to Newfoundland on compassionate leave, January 1919. Veterans Affairs, *The Royal Newfoundland Regiment*. Regimental History Ledger Sheets, 1, Officers; Reserve Officers; Attached Officers.

36 Lieutenant S. Robertson. Robertson was wounded at Beaumont Hamel and carried back across No Man's Land by a brother officer, Captain J. A. Ledingham. No other details on Robertson's service record appear in published sources. See Nicholson, *The Fighting Newfoundlander*, 275.

37 Lieutenant Eric S. Ayre was a grandson of C.R. Ayre, founder of a well-known and respected St John's firm. He was one of two officers appointed in 1915 to the Reserve Force Committee, a committee of the Newfoundland Patriotic Association charged with responsibility for enrolling and training drafts for the Newfoundland Regiment. It later changed its name to the Standing Committee on Military Organization. He was promoted Captain and put in temporary command of D Company, a

draught that was to leave for England in March 1915. He accompanied E Company to England in April 1915. He lost his life at Beaumont Hamel, on 1 July 1916, one of four members of his family to take the supreme sacrifice, all on that same day. Nicholson, *The Fighting Newfoundlander*, 197, 206, 209, 274.

38 Robert Templeton Smith.

39 Editor of the *Daily News*.

40 George Hayward Taylor. Enlisted 2 September 1914; Sergeant, 21 September 1914; Colour Sergeant, 3 October 1914; Company Quartermaster Sergeant, 29 October 1914; Company Sergeant Major, 31 July 1915; Second Lieutenant, 9 October 1915; BMEF, 14 November 1915; BEF, 14 March 1916; killed in action, Beaumont Hamel, 1 July 1916. Cramm, *The First Five Hundred*, 296

41 Steele had spent some time in Vancouver with the firm of Henry Birks and Sons. He may have met R. Dalton during this period. See *Dictionary of Canadian Biography*, 14: 966–7.

42 Wife of former governor of Newfoundland. Nicholson, *The Fighting Newfoundlander*, 149–50.

43 Herbert Rendell had been an employee of C.F. Bennett & Co., St. John's, *St. John's City Directory 1919*, Wounded in October at Suvla; commanded a rearguard in the firing line during the evacuation.

44 Owen's brother, Victor John Steele, born 31 December 1897.

45 Walter Martin Greene. Enlisted 2 September 1914; Lance Corporal, 21 September 1914; Corporal, 13 November 1914; Provost Sergeant, 23 April 1915; BMEF, 20 August 1915; awarded DCM, 24 January 1916; BEF, 14 March 1916; 2nd Lieutenant, 5 June 1916; wounded, Somme Raid, 28 June 1916; invalided to England, 5 July 1916; returned BEF, 4 May 1917; Lieutenant, 1 November 1917; killed in action, Marcoing, 20 November 1917. Cramm, *The First Five Hundred*, 189.

46 Major T.M. Drew, of the Leicestershire Regiment, became second-in-command in July 1915. Because of ill-health he returned to England in 1916, just before Beaumont Hamel. Nicholson, *The Fighting Newfoundlander*, 151, 250.

47 Major C.W. Whitaker, of the Liverpool Regiment, joined the Newfoundland Regiment in July 1915. He became Lieutenant-Colonel of the regiment in July 1916, a position which he held until January 1918. Nicholson, *The Fighting Newfoundlander*, 151, 287, 440.

48 Reginald S. Rowsell. Commissioned Lieutenant, 24 September 1914; Captain, 28 July 1915; BMEF, 20 August 1915; BEF, 14 March 1916; wounded, Beaumont Hamel, 1 July 1916; invalided to England, 3 July 1916; attached to depot, Ayr, 23 August 1916; awarded Military Cross 1 January 1917; returned to BEF, 9 January 1917; killed in action, Monchy, 14 April 1917. Cramm, *The First Five Hundred*, 128.

CHAPTER THREE

1 Lind, *The Letters*, 65; Nicholson, *The Fighting Newfoundlander*, 151–2.
2 There are two different versions of this speech. Nicholson, in *The Fighting Newfoundlander,* 154, adds after "prepared," "and sharpen those bayonets for the Turks" but does not indicate his source. I have used Lind, believing his account to be more immediate and therefore more accurate. Lind, *The Letters of Mayo Lind*, 69.
3 Lind, *The Letters*, 70. While at Aldershot King George V inspected the regiment, for a second time (Nicholson, *The Fighting Newfoundlander*, 154). None of this detail appears in Steele's account. It may have been that he was too busy to keep a lengthy account of his activities. The regimental history mentions that the last few days at Aldershot were particularly hectic for officers. Nicholson, *The Fighting Newfoundlander*, 155.
4 Lind, *The Letters*, 72.
5 Nicholson, *The Fighting Newfoundlander*, 156–7.
6 Ibid., 154.
7 Lind, *The Letters*, 77–80.
8 Ibid., 81.
9 Nicholson, *The Fighting Newfoundlander*, 159.
10 National Archives of Canada (NAC) MG31 G19. Nicholson, Gerald William Lingen. Correspondence 1903–1979. Volume 1, Correspondence 1961–1963. Bob Mills to Mr. G.W.L. Nicholson, 4 August 1963.
11 Ziegler, *Omdurman*, 19–20.
12 Lind, *The Letters*, 82.
13 NAC MG31 G19 Nicholson Correspondence, 1, Mills to Nicholson, 4 August 1963.
14 Lind, *The Letters*, 84–5.
15 Lind, *The Letters*, 69.
16. Farnborough, only 4 kms distant from Aldershot, was an air station. Lind, *The Letters*, 68.
17 Cyril C. Duley of St John's, Newfoundland. Enlisted 2 January 1915; commissioned in the field on 27 October 1916; wounded twice in December 1916 and hospitalized in England; returned to Newfoundland on 5 March 1917; on 3 September 1918 found permanently unfit due to bronchitis and wounds; appointed adjutant of the regiment on 16 March 1919; retired the following month on 30 April 1919; awarded the MBE on 3 June 1919. The Royal Newfoundland Regiment, Regimental History Ledger Sheets. Vol. 1. Officers, Reserve Officers, Attached Officers.
18 Lieutenant Kevin J. Keegan. Battalion signalling officer, one of "the men who saved Monchy." During an action on 14 April 1917 at Monchy-le-Preux, the Newfoundland Regiment, along with other units of the 88th

Brigade, found themselves under a strong German counter-attack. Nearly 300 of the regiment were killed, wounded, or missing in the action. While the brigade was under attack Lt. Col. Forbes-Robertson and a small party of about a dozen others from battalion headquarters rushed forward to a position overlooking the German advance. From a parapet in a disused trench the small band, reduced to ten, held off the German advance for four hours. That valiant defence was vital to the British effort in that sector. Once the detachment was relieved, Lieutenant Keegan went out with two of his men to bring back five wounded Newfoundlanders from the battleground. Keegan was awarded the Military Cross for his role at Monchy. By October 1917 Keegan was Captain in charge of "C" Company. He was wounded at Broembeek on 9 October, for which he received a bar to his M.C. Nicholson, *The Fighting Newfoundlander*, 351–3, 355–6, 360, 392, 395, 399.

19 Nicholson, *The Fighting Newfoundlander*, 155. The *Megantic* was a 16,000 ton ship that had been part of the convoy crossing the Atlantic the previous October that included the *Florizel*.

20 Nicholson, *The Fighting Newfoundlander*, 158. The Warwickshires were composed of veteran soldiers of campaigns in India and Africa. Some of them had been wounded while serving with the BEF in 1914 and were no longer considered fit for front line duty.

21 Gerald Harvey: Enlisted 10 September 1914; Second Lieutenant, 6 April 1915; BMEF, 20 August 1915; wounded, Suvla, 3 October 1915; invalided to England, 6 October 1915; Lieutenant, 15 October, 1915; attached to depot, Ayr, 11 April 1916; returned to Newfoundland on furlough, 17 April 1916; returned to U.K., 27 July 1916; attached to Royal Flying Corps, 26 September 1916; rejoined Second Battalion, 7 March 1917; BEF, 4 May 1917; wounded, Belgium, 7 October 1917; invalided to England, 23 October 1917; returned to Newfoundland, 30 January 1919; retired 25 February 1919. Cramm, *The First Five Hundred*, 195.

22 Henry S. Windeler eventually became a major and received the Military Cross for action at Bellenglise while temporarily attached to the 46th Battalion, Machine Gun Corps. Nicholson, *The Fighting Newfoundlander*, 575.

CHAPTER FOUR

1 Stacey and Edwards, *Memoirs of a Blue Puttee*.
2 Among others one can point to James, *Gallipoli*; and Moorehead, *Gallipoli*.
3 Moorehead, *Gallipoli*, 301–4.
4 Keegan, *The First World War*, 248, 234–48; see also James, *Gallipoli*, 261–2.

5 Major R.H. Tait, M.C., "With the Regiment at Gallipoli," PANL, *The Veteran*, 1: no. 2 (April 1921): 38-52 passim; Lind, *The Letters*, 86–9.

6 The 29th Division was organized in January 1915 and was the last infantry division to be formed from regular battalions of the British army. Of its twelve battalions, formed into three brigades, the 86th, 87th, and 88th, eleven were drawn from India, Burma, China, and Mauritius. The twelfth was an Edinburgh Territorial unit, the 1st Battalion of the 5th Royal Scots. It suffered heavy losses in the early fighting in Gallipoli and was replaced by the Newfoundland Regiment, the only colonial unit in the division. Nicholson, *The Fighting Newfoundlander*, 168–9.

7 Gallishaw, *Trenching at Gallipoli*, 104–5.

8 Lind, *The Letters*, 101, 105.

9 James D. Atwater, "Echoes and Voices," *Smithsonian* (November 1987): 208.

10 Cramm, *The First Five Hundred*, 42–45; PANL, Tait, "With the Regiment at Gallipoli," 38.

11 Nicholson, *The Fighting Newfoundlander*, 179–81.

12 James, *Gallipoli*, 334–6.

13 The 4th Battalion of the Worcestershire Regiment was one of the regular battalions of the 88th Brigade to which the Newfoundland Regiment belonged.

14 James, *Gallipoli*, 334–6.

15 Nicholson provides lively detail of the storms and their effect on the regiment. Nicholson, *The Fighting Newfoundlander*, 181–4.

16 The origins of the Cunard Line go back to 1840 with the establishment of the British and North American Royal Mail Steam Packet Company. The name was changed to the Cunard Steamship Company Limited in 1878. In 1934 the Cunard and White Star Lines merged to form the Cunard-White Star Limited. Fifteen years later, in 1949, the name changed again to the Cunard Steamship Company Limited. Since 1962 it has been known as the Cunard Line Limited. The 7,907 ton *Ausonia*, formerly the *Tortona*, was taken over by Cunard with the Thomson Line in 1911. It was torpedoed and sunk by gunfire in 1918 with the loss of forty-four lives. A second ship, the HMS *Ausonia*, was launched in 1922 and served for twenty years with the Royal Navy as fleet repair ship. www.theshipslist.com/ships/lines/cunard.html.

17 Private Hugh McWhorter, from the Bay of Islands, the regiment's first casualty, was killed 22 September 1915, at age 21. He is buried at Hill 10 Cemetery, Suvla Bay. See photo in C.S. Frost collection, PANL. *The Letters of Mayo Lind*, 87; Nicholson, *The Fighting Newfoundlander*, 171.

18 Sam Lodge. Enlisted, 2 September 1914; BMEF, 20 August 1915; killed in action, Suvla Bay, 1 October 1915. Cramm, *The First Five Hundred*, 227.

19 Lance Corporal Walter Tucker died twelve days later. Lind, *Letters*, 100.

20 Thomas Crawford Gowans of St. John's, Newfoundland, Regimental
   number 1177; enlisted 21 February 1915; wounded, Suvla Bay, 16 October 1915; hospitalized until 11 February 1916; posted to Headquarters,
   18 July 1917; embarked for Newfoundland, 15 July 1917; Lance
   Corporal, 10 June 1919; Corporal, 1 December 1919. Veterans Affairs,
   Canada. The Royal Newfoundland Regiment. Regimental History,
   Ledger Sheets Volume 3.

21 Lieutenant Cyril Carter. Nicholson refers to the same incident in *The
   Fighting Newfoundlander*, 181.

22 James Robert Steele, younger brother of Lieutenant Owen Steele. Enlisted, 11 January 1915; sailed on ss *Stephano*, 20 March 1915; Corporal
   22 February 1915; Sergeant, 14 November 1915; BMEF, 20 August 1915;
   BEF, 14 March 1916; wounded Beaumont Hamel, 1 July 1916; evacuated
   to Wandsworth Hospital, England; attached to "E" Company Depot
   Cadet College, Trinity, Cambridge, 16 July 1917; Second Lieutenant,
   29 May 1917; BEF, joined unit in field, 22 December 1917; wounded,
   Ypres, 26 March 1918; evacuated to Wandsworth Hospital, 4 April
   1918; posted to "H" Company Second Battalion, Winchester, 20 May
   1918; posted to Newfoundland for special duty, 2 July 1918. Compiled
   by James Steele, nephew of Owen, from interviews and extracts from
   letters home.

23 A Newfoundland dish made from hard bread soaked overnight and
   warmed just to the boiling point. It is normally served with salt cod
   that has been soaked in several changes of water and boiled until
   cooked. In the more traditional form of this Newfoundland dish,
   salt pork bits, fried and rendered out, are poured over the fish and
   brewis.

24 Ernest St. Claire Churchill. Enlisted, 8 September 1914; Lance Corporal,
   21 September 1914; Corporal, 3 October 1914; Sergeant, 21 April 1915;
   Second Lieutenant, 16 August 1915; BMEF, 20 August 1915; evacuated
   Suvla, sick, 6 November 1915; invalided to England, 4 April 1916;
   attached to depot, Ayr, 4 July 1916; returned to Newfoundland, 23
   August 1916; arrived in U.K., 30 April 1917; BEF, 4 May 1917; evacuated
   to hospital, 21 June 1917; invalided to England, 8 July 1917; attached to
   depot, Ayr, 29 November 1917; returned to BEF, 5 January 1918;
   wounded, Belgium, 28 March 1918; invalided to England, 31 March
   1918; returned to Newfoundland, 22 May 1919; retired 5 July 1919.
   Cramm, *The First Five Hundred*, 155.

25 Lieutenant J.J. Donnelly. Active in the Catholic Cadet Corps, St. John's,
   in the pre-war years; Lieutenant, Gallipoli; winner of Newfoundland
   Regiment's first military cross for the part he played in capturing and
   holding Caribou Hill, Suvla; commanded "C" Company at Gueudecourt;
   killed in action at the capture of Hilt Trench, Gueudecourt. Compiled

by James Steele from interviews and extracts from letters home. See also Nicholson, *The Fighting Newfoundlander*, 95, 179–80, 244, 310.

26 H.H.A. Ross. Wounded while leading reinforcement party on Caribou Hill, Suvla, October 1915; wounded again in France; appointed Adjutant Forestry Corps, 1916. Compiled by James Steele from interviews and extracts from letters home. See also Nicholson, *The Fighting Newfoundlander*, 180.

27 Edward Joseph Murphy. Enlisted 4 September 1914; BMEF, 20 August 1915; evacuated to England, 16 November 1915; BEF, 28 March 1916; killed in action, Beaumont Hamel, 1 July 1916. Cramm, *The First Five Hundred*, 249.

28 Major-General Sir Beauvoir de Lisle came from France to take command of the 29th Division on 4 June 1915. James, *Gallipoli*, 218.

29 Brigadier General D.E. Caley was commander of the 88th Brigade of which the Newfoundland Regiment became a part in 1915. Nicholson, *The Fighting Newfoundlander*, 169.

30 For their part in the capture of "Caribou Hill" Lieutenant Donnelly and Lance-Corporal Fred Snow of St John's received the Military Cross; Sergeant W.M. Greene of Avondale and Private R.E. Hynes of Indian Island, Fogo, won Distinguished Conduct Medals. Nicholson, *The Fighting Newfoundlander*, 180.

31 Lieutenant John Grierson Bethune of Edmonton, Alberta, Regimental Number 845. Enlisted, 29 December 1914; promoted Sergeant, 10 July 1915; commissioned 2nd Lieutenant, 14 September 1916; Lieutenant, 25 January 1918; served with the regiment until September 1917; as a result of a debility was released, eventually returned to Newfoundland; discharged, 1 May 1918 as medically unfit. Veterans Affairs Canada, The Royal Newfoundland Regiment, Regimental History Ledger Sheets, vol. 1. Officers, Reserve Officers, Attached Officers.

32 The 34th Brigade, under the command of Brigadier-General W.H. Sitwell, captured Chocolate Hill, 7 August 1915. James, *Gallipoli*, 281.

33 Major T.M. Drew was from the Leicestershire Regiment and became Lieutenant-Colonel Burton's second-in-command. When Burton was wounded at Gallipoli Drew took over command of the regiment. Succumbed to fever in Gallipoli and was hospitalized; rejoined the regiment in April 1916; took temporary command of the regiment while Hadow was in England to receive the CMG for his services in Gallipoli; replaced by Major James Forbes-Robertson on medical grounds, 15 June just before the battle of Beaumont Hamel. Nicholson, *The Fighting Newfoundlander*, 151, 178–9, 181, 244, 250.

34 James Joseph Hynes. Killed 18 November 1915. Nicholson, *The Fighting Newfoundlander*, 557.

35 Actually Southside Hills. This refers to hills that dominate the southern side of the harbour of St John's. Haynes lived on the narrow strip of land that extends from the hills to the harbour.

36 Richard E. Hynes was awarded the Distinguished Conduct Medal, second in importance to the Victoria Cross for noncommissioned officers and men, for "conspicuous gallantry" in the action at Caribou Hill, Gallipoli, on the night of 4/5 November 1915. It was the regiment's first significant engagement. Lance Corporal Hynes was killed at Beaumont Hamel. Nicholson, *The Fighting Newfoundlander*, 179–80, 244–5, 557; Cramm, *The First Five Hundred*, 41.

37 Richard A. Shortall. Enlisted, 5 September 1914; Lance Corporal, 21 September 1914; Corporal, 11 March 1915; Second Lieutenant, 22 April 1915; BMEF, 20 August 1915; wounded, Suvla, 26 November 1915; admitted hospital, Mudros, 30 November 1915; Lieutenant, 1 January 1916; transferred to Malta, 18 January 1916; rejoined Battalion, Suez, 7 March 1916; BEF, 14 March 1916; killed in action, Beaumont Hamel, 1 July 1916. Cramm, *The First Five Hundred*, 283.

38 Cecil Bayly Clift. Enlisted, 7 September 1914; Lance Corporal, 26 April 1915; Second Lieutenant, 29 June 1915; BMEF, 20 August 1915; Lieutenant, 1 January 1916; BEF, 14 March 1916; evacuated to hospital, 20 April 1916; rejoined Battalion, 14 July 1916; killed in action, Gueudecourt, 12 October 1916. Cramm, *The First Five Hundred*, 159.

39 Captain A.E. Bernard. Before the war Bernard was one of the masters at Bishop Feild College, an Anglican academy in St John's. He was in England when the war broke out and, with Arthur Raley, another Feildian master, returned to St John's to enlist. He was one of the first company commanders appointed to the battalion during its early training at Pleasantville, St John's. Bernard's rendition of the *La Marseillaise* in the concert referred to by Steele was especially popular and led to a request for a repeat performance. He was promoted Major and took over the post of second-in-command of the battalion in 1916. He served with the regiment for the duration of the war, ending with the rank of Lieutenant-Colonel. He received the DSO for action at Cambrai as well as the MC for distinguished service. Nicholson, *The Fighting Newfoundlander*, 112, 119–20, 326, 504–7, 573–4.

40 Robert Bruce Reid. Enlisted 8 September 1914; Lance Corporal, 10 August 1915; Second Lieutenant, 16 August 1915; BMEF, 20 August 1915; BEF, 14 May 1916; killed in action, Beaumont Hamel, 1 July 1916. Cramm, *The First Five Hundred*, 271; Nicholson, *The Fighting Newfoundler*, 275.

41 Alexander Frew. Attached to the 1st Newfoundland Regiment as Medical Officer, 17 August 1915; struck off strength, 1 September 1916. Veterans Affairs, Canada, The Royal Newfoundland Regiment, Regi-

mental History Ledger Sheets, vol. 1. Officers, Reserve Officers, Attached Officers.

42 Henry Stanton Windeler of St John's. Appointed Lieutenant 26 January 1915; Captain, 14 November 1916, and temporary Major 1 January 1917. He was awarded the Military Cross for action at Bellenglise while temporarily attached to the 46th Battalion, Machine Gun Corps. He retired with the honorary rank of Major on demobilization on 16 June 1919. Veterans Affairs Canada. The Royal Newfoundland Regiment, Regimental History Ledger Sheets; Nicholson, *The Fighting Newfoundlander*, 575.

43 Rupert Wilfred Bartlett. Enlisted, 2 September 1915; Lance Corporal, 21 September 1914; Corporal, 21 April 1915; Second Lieutenant, 22 April 1915; Lieutenant, 16 October 1915; BMEF, 14 November 1915; BEF, 14 March 1916; wounded, Gueudecourt, 12 October 1916; evacuated to England, sick, 15 October 1916; returned to unit 12 March 1917; awarded Military Cross, 18 June 1917; awarded bar to Military Cross, 26 September 1917; killed in action, Marcoing, 30 November 1917; foreign decoration, Order of the Crown of Italy, Cavalier, 29 November 1917. Cramm, *The First Five Hundred*, 137.

44 Charles St. Clair Strong. Enlisted, 2 September 1914; Sergeant, 21 September 1914; Color Sergeant, 3 October 1914; Company Sergeant Major, 6 May 1915; wounded, Somme Raid, 28 June 1916; Second Lieutenant, 9 October 1915; Lieutenant, 9 October 1916; Captain, 16 August 1917; wounded, Neuve Eglise, 12 April 1918; died of wounds, 13 April 1918. Cramm, *The First Five Hundred*, 295.

CHAPTER FIVE

1 Terraine, *The First World War 1914–18*, 86.

2 James, *Gallipoli*, 338.

3 NAC MG31 G19. Fighting Newfoundlander Correspondence: 1961–1963. Extract from a letter written by Lieutenant-Colonel A. Hadow, 20 December 1915.

4 James, *Gallipoli*, 344.

5 Arthur Lovell Hadow joined the Norfolk Regiment in 1898, just before his twenty-first birthday. He spent 1904 in Tibet as part of the escort to Colonel Francis Younghusband's expedition and marched with him to Lhasa. On return he joined his regiment in South Africa where he served with the mounted infantry for six months. He found garrison duty particularly galling and requested transfer to the Egyptian Army, then held in high regard following Lord Kitchener's tour of duty there as its Commander-in-Chief. Hadow was in charge of an isolated outpost on the Congo northern border for seven months, after which he transferred to the Sudan Civil Administration as district commissioner. He remained in the Sudan until

1915 when he was sent to Gallipoli for two months. There he served as Staff Captain of the 88th Brigade and then as Brigade Major of the 34th Brigade of the 11th Division. His appointment with the Newfoundland Regiment took effect on 6 December 1915 on the authority of the Army Council but without the prior approval of the Newfoundland government. See Nicholson, *The Fighting Newfoundlander*, 184–5 and also NAC CBC, Radio: Flanders Fields ISN 117216. CO7721 (3) Flanders Fields: [background interview] Hadow, A.J. – Royal Newfoundland Regiment – Interview May 1964.

6   NAC CBC, Radio: Flanders Fields ISN 117216. CO7721 (3) Hadow, A.J. – Royal Newfoundland Regiment, Interview May 1964.

7   James, *Gallipoli*, 346. This detail does not appear in the typescript of the letters nor in the diary. James seems to have consulted Ella Steele directly, as his Notes on Sources state that the diaries of Lt. O.W. Steele were in possession of his sister, Miss E. Steele. She may have provided additional detail based on material not included in the typescript.

8   Bombing refers to grenades. The bombing officer was responsible for instructing men in the battalion in the use of the grenade in assault. Griffith, *Battle Tactics of the Western Front*, 69–70.

9   Corporal John J. Moakler, Regimental Number 782. Enlisted on 18 December 1914; served with the regiment at Suvla; wounded in the thigh, 6 January 1916; hospitalized for wound and recurring ailments until December 1916; promoted Lance Corporal, 17 September 1917, Acting Corporal, 4 November 1918, confirmed Corporal, 4 November 1918. Veterans Affairs Canada, The Royal Newfoundland Regiment. Regimental History, Ledger Sheets, vol. 2.

10 Navvy, plural navvies: British term for a labourer, especially one employed in the construction of excavation projects.

11 L. Cpl.Thomas Cook. Killed 3 December 1917. Nicholson, *The Fighting Newfoundlander*, 550.

12 Private George Simms. Killed 30 December 1915. Nicholson, *The Fighting Newfoundlander*, 568.

13 When war broke out in August 1914 the German battle-cruiser *Goeben* and her escort, the light cruiser *Breslau*, then in the Mediterranean, escaped pursuing British squadrons and gained entry to Dardanelles, eventually anchoring off the Golden Horn. Turkey's decision to grant the two German ships sanctuary effectively allied it to the Central Powers for the rest of the war. James, *Gallipoli*, 8–10.

14 The reference to "Asiatic" refers to the location of guns on the Asiatic shore of the Dardanelles where the 3rd and 11th Divisions of the Turkish Army were located. James, *Gallipoli*, 252.

15 This may be Rupert Wilfred Bartlett.

16 The "X" Beach was located on the west coast of Gallipoli Peninsula, approximately three-quarters of a mile north of Tekke Burnu; "W" Beach

was between Tekke Burnu and Cape Helles, referred to as Lancashire Landing; "V" Beach was located a mile east of "W" Beach. The Helles landings were made by the 29th Division on these three beaches. James, *Gallipoli*, 88; Nicholson, *The Fighting Newfoundlander*, 190, map of three beaches, 191.

17 John Clift. Enlisted 7 September 1914; appointed Lance Corporal, 26 April 1915; struck off strength and commissioned with Cameron Highlanders, 8 April 1915; transferred from Cameron Highlanders to Newfoundland Regiment as Captain, 26 October 1917; BEF, 10 November 1917; awarded Military Cross, 16 September 1918; returned to Newfoundland where retired, 25 February 1919. See Cramm, *The First Five Hundred*, 159.

18 Lieut-General Sir James Wilton O'Dowda, KCB, CSI, CMG (1871–1961). Entered Royal West Kent Regiment in 1891, rising to the rank of major in 1909. Served as Lt-Col of the Royal Dublin Fusiliers in 1915 and promoted to the rank of Brigadier-General that same year. He was promoted to the rank of Maj.-Gen. in 1925 and Lt.-Gen. in 1931. Served in the North-West Frontier, India, 1897–98, and in the First World War in Egypt, Gallipoli, and Mesopotamia, and in the Afghan in 1919. He served thereafter in various command positions in Indian and in Britain. He retired in 1934. *Who Was Who 1961–1970*, 6: 850.

19 Possibly the Brigade Major. No other identification.

20 For further detail see Nicholson, *The Fighting Newfoundlander*, 192.

21 Gully Ravine, Cape Helles. The Newfoundland Regiment, as part of the 29th Division, was sent back to the Peninsula from Imbros, where it joined the 88th Brigade reserve. There it furnished daily fatigue parties to work on the repair of defences along a portion of Fusilier Street, a reserve line that ran from the eastern end of the peninsula to Gully Ravine, opposite "Y" Beach on the Aegean side of the peninsula. See Nicholson, *The Fighting Newfoundlander*, 189–90, map, 191.

22 Thomas Mouland. Enlisted, 11 September 1914; BMEF; wounded Cape Helles, 7 January 1916; admitted hospital, Malta, 13 January 1916; invalided to England, 12 May 1916; admitted St. Dunstan's Hospital for the Blind, 19 August 1916; discharged, United Kingdom, medically unfit, 15 June 1917. Cramm, *The First Five Hundred*, 241.

23 Private Robert Morris. Killed 7 January 1916. Nicholson, *The Fighting Newfoundlander*, 561.

24 Private Michael Mansfield. Killed 13 April 1918 at Bailleul. Nicholson, *The Fighting Newfoundlander*, 450–9, 560.

25 Neither Dunlop nor Yates have been identified. They were either from divisional or brigade staff.

26 Major-General F.S. Maude, Commander of the 29th Division. See Nicholson, *The Fighting Newfoundlander*, 192.

27 A mine boom, a floating wooden barrier used to prevent passage of float-
    ing mines.
28 Burton was hit in the hand at Suvla Bay on 30 October 1915 and had to
    be evacuated. Nicholson, *The Fighting Newfoundlander*, 178.

CHAPTER SIX

1 NAC, ISN 117216, CO7721 CBC. Flanders Fields Hadow Interview.
2 Nicholson, *The Fighting Newfoundlander*, 229.
3 See Warren, "Womens' Patriotic Association."
4 Nicholson, *The Fighting Newfoundlander*, 32.
5 Burton was the regiment's former C.O.
6 The Anglo-American Telegraph Company installed the first successful
   trans-Atlantic cable in 1866. It amalgamated with the New York,
   Newfoundland and London Telegraph Company in 1873 and obtained
   a fifty-year monopoly on landing cables in Newfoundland, that lasted
   until 1904. By the end of the nineteenth century telegraphic service
   was extended to the major communities on the island and by 1904 to
   those in Labrador. When the Anglo-American lease ran out, further
   construction of wireless telegraphy stations along the Newfoundland
   and Labrador shorelines, known as coastal stations, was undertaken by
   another company, the Canadian Marconi Company. During the war
   the Anglo-American provided the only trans-Atlantic telegraphic link
   between England and Newfoundland. Poole, *Encyclopedia of Newfoundland and Labrador*, 346–7; Harding "Wireless in Newfoundland,
   1901–1933," 15.
7 Cluny Macpherson (1879–1966) was born in St John's. He was educated
   at Methodist College, St. John's, and at McGill University. On graduation from medical school he was one of the first Newfoundland doctors
   to be associated with Sir Wilfred Grenfell and the Royal National Mission to Deep Sea Fishermen. He developed an early association with the
   St John Ambulance. On the outbreak of war he enlisted in the Newfoundland Regiment and served as its principal medical officer and was
   seconded to the War Office to develop one of the earliest prototypes of
   the gas mask. He was invalided home following an injury in Egypt
   where he served as director of Medical Services for the Militia. He left
   the service in 1919 with the rank of Lieutenant-Colonel and continued
   to play an active role in the Newfoundland medical profession until his
   death. Smallwood and Pitt, *Encyclopedia of Newfoundland and
   Labrador*, 3: 424–5.
8 The regulation overcoat for officers.
9 Presumably the same Wilson identified earlier as Commanding Royal
   Engineers with the 88th Brigade.

10 These are members of the staff of 88th Brigade to which the Newfoundland Regiment belonged. None of these individuals is identified in regimental sources.

11 This is the Kodak camera that his family were so eager to recover when Steele was killed. The photographs were deposited in the Centre for Newfoundland Studies Archives at Memorial University of Newfoundland by James Steele, nephew of Owen and son of James.

12 Later Company Sergeant Major, Maurice Power. Nicholson, *The Fighting Newfoundlander*, 577.

13 Sergeant Harold Mitchell of St. John's; Regimental Number 828; enlisted 21 December 1914; promoted Lance Corporal, 19 April 1915; Corporal, 26 May 1915, Sergeant, 26 September 1915; became ill in Suvla, December 1915, later diagnosed as typhoid and enteric fever; was transferred to hospital in England from which he was later home to Newfoundland; discharged as medically unfit, 25 April 1917. Veterans Affairs Canada The Royal Newfoundland Regiment, Regimental History Ledger Sheets, vol. 5.

14 Richard E. Hynes, later Lance Corporal; awarded Distinguished Conduct Medal for action at Caribou Hill. Nicholson, *The Fighting Newfoundlander*, 575.

15 John Fitzgerald. Enlisted, 8 September 1914; BMEF, 20 August 1915; killed in action, Suvla, 1 December 1915; mentioned in despatches, *London Gazette*, 11 July 1916. Cramm, *The First Five Hundred*, 179; Nicholson, *The Fighting Newfoundlander*, 184. A Red Cross man was a stretcher bearer.

16 At the beginning of the war each battalion had two Vickers machine guns, deployed under the direction of a Lieutenant Colonel. The water-cooled Vickers was replaced by the air-cooled Lewis machine gun in 1915. Although lighter than the Vickers, the Lewis was still too heavy – at 12 kg – for portable use. It could fire up to 500 rounds per minute, although normally was fired in short bursts. As these weapons became more plentiful, their direction devolved upon companies and eventually platoons. This meant that company officers like Steele had to complete the requisite course for their employment. Griffith, *Battle Tactics of the Western Front*, 21–2; www.greatwar.org/weapons/machineguns.htm.

17 The Suez Canal was opened in November 1869. When the Egyptian Khedive defaulted on his loans in 1879 the British government bought shares in the canal. A revolt of the Egyptian Army in 1882, which threatened Britain's vital access to the canal and the route to India, led to an eventual British occupation of Egypt that was to last until 1952. In the period before the Great War British defence planners spent considerable energy trying to find ways of defending the canal against a possible Turkish attack. By January 1915, 70,000 troops, Australian, British territorials,

Indian, and New Zealand, guarded the canal. The Turkish attack in 1915 failed and the threat to the canal diminished thereafter. Keegan, *The First World War*, 219–21.

18 Herbert Henry Asquith, first Earl of Oxford and Asquith (1852–1928). At the time Asquith was prime minister of the coalition government formed in May 1915. He and his principal liberal colleagues resigned from the government in December 1916 and he was succeeded by Lloyd George. Asquith remained in parliament as leader of the opposition until he was defeated in the election of 1918. He returned to parliament briefly in 1920 and resigned as leader of the Liberal Party in 1926. *The Dictionary of National Biography*, 2: 14–5.

19 The machine gun had a two person gun crew. As it weighed 12 kilograms it required two men to mount it on the tripod and to operate it.

20 Baths refer to mobile bath units set up at central locations by the military for routine showering and for laundering of articles of clothing.

21 Clifford Jupp. Enlisted, 5 September 1914; Lance Corporal, 12 June 1915; BMEF, 20 August 1915; Corporal, 26 September 1915; Sergeant, 14 November 1915; BEF, 14 March 1916; Acting Company Quartermaster Sergeant, 29 May 1916; Second Lieutenant, 11 June 1916; killed in action, Beaumont Hamel, 1 July 1916. Cramm, *The First Five Hundred* , 213.

22 Nicholson, *The Fighting Newfoundlander*, 233.

CHAPTER SEVEN

1 Terraine, *The First World War, 1914–18*, 95

2 Edmund Henry Hynman, 1st Viscount Allenby (1861–1936) British Field Marshal. Born in Felixstowe, England; educated at Haileybury College and the Royal Military College, Sandhurst. Between 1884 and 1902 served with the Inniskilling Dragoons in Bechuanaland and Zululand and later in the Boer War. He was appointed Inspector of Cavalry in 1910. During World War I he commanded the British cavalry in France. He took over command of the Fifth Army Corps in 1915. For his highly successful command of the Third Army at the Second Battle of Ypres and in the capture of Vimy Ridge he was promoted to the rank of general and made a Knight Commander of the Bath. He took over command of the Egyptian Expeditionary Force in 1917 and led the successful offensive against Turkish armies in the Middle East. He captured Jerusalem in December 1917, followed by victories against Turkish forces at Megiddo in September 1918 and the capture of Damascus in 1918 that forced Turkey to capitulate. He was promoted to the rank of Field Marshal and made a Viscount for his service in the Middle East. He remained on in Egypt after the war as British High Commissioner. He was appointed

Lord Rector of the University of Edinburgh in 1936, the year he died. *Encarta Encyclopedia Online Deluxe.*

3 Nicholson, *The Fighting Newfoundlander*, 237.

4 Lyn Macdonald, *Somme*, 18.

5 Also "swinging the lead," a slang expression meaning to feign illness in order to avoid work.

6 A metal pipe filled with explosives and used to clear a path through barbed wire or to detonate mines.

7 Nicholson, *The Fighting Newfoundlander*, 259–60; Raley, "The Deadly First of July Drive," Smallwood and Pitt, *Encyclopedia of Newfoundland and Labrador*, 6: 541–2.

8 PANL Aa-2-4. War Diaries. Microfilm B-2302. Entry 26 June 1916.

9 Steele, "A Few Notes."

10 Lind, *Letters*, 157.

11 Single shot artillery fire used to determine range of a target.

12 PANL Aa-2-4. War Diaries. Entries 26 and 27 June 1916; PANL, Major R. Butler, "Raid of June, 1916"; Cramm, *The First Five Hundred*, 55–6.

13 Nicholson, *The Fighting Newfoundlander,* 259

14 Steele, "A Few Notes."

15 Major (later Lieutenant Colonel) James Forbes-Robertson of the 1st Battalion, the Border Regiment (of the neighbouring 87th Brigade) joined the regiment as its new Second-in-Command on 15 June 1916, replacing the ailing Major T.M. Drew. At Beaumont Hamel, Forbes-Robertson was in charge of the ten per cent kept back as a nucleus on which the unit might be rebuilt in the event of heavy casualties of which Steele was a part. Following the battle Forbes-Robertson instituted a strict regime of training from 5:30 a.m. until 7:30 p.m. whose object was to keep the men's minds off the terrible slaughter at Beaumont Hamel. While it was an unpopular measure, it worked. Nicholson recorded that out of the ruins of Beaumont Hamel emerged a new battalion with a strong *esprit de corps*. Forbes-Robertson replaced a worn-out Lieutenant-Colonel Hadow on 27 November 1916. Five months later, in April 1917, at Monchy-le-Preux, the regiment was nearly annihilated in an unexpected German counter-attack on the 88th Brigade as it advanced against the German line. Lieutenant-Colonel Forbes-Robertson and a handful of Newfoundlanders held off the German advance, thereby preventing the recapture of Monchy-le-Preux. For his role in this heroic defence Forbes-Robertson was awarded the DSO. When Hadow returned to the regiment in May 1917 Forbes-Robertson reverted to his former rank. In August 1917 he took over command of the 16th Battalion, the Middlesex Regiment. Nicholson referred to it as a "well deserved promotion." Forbes-Robertson was awarded the MC and, notably, the VC, while serving with the Border Regiment. Nicholson,

The Fighting Newfoundlander, 250–1, 262, 277, 321, 345–6, 351–7, 359, 378, 387–8, and Appendix B.

16 Nicholson, *The Fighting Newfoundlander*, 261–3.

17 Ralph Barnes Herder. Enlisted, 2 September 1914; Lance Corporal, 26 July 1915; BMEF, 20 August 1916; 2nd Lieutenant, 1 July 1916; wounded Beaumont Hamel, 1 July 1916; invalided to England; returned to BEF, 27 October 1916; wounded, Monchy, 14 April 1917; repatriated to Newfoundland; Lieutenant, 1 January 1918; retired, 30 June 1918. Cramm, *The First Five Hundred*, 198.

18 Wilfred D. Ayre was a grandson of C.R. Ayre, founder of one of St John's mercantile firms and was related to Captain Eric Ayre and Second Lieutenant Gerald W. Ayre, both officers with the regiment. He was killed at Beaumont Hamel. Nicholson, *The Fighting Newfoundlander*, 275.

19 A communications trench running from the south side of Auchonvillers to the sunken road which connected that village with Hamel. The Beaumont Hamel map, inserted between 266–7 of Nicholson, gives a clear rendition of the location. Nicholson, *The Fighting Newfoundlander*, 241, 253, 263.

20 Peter J. Cashin (1890–1977), son of Sir Michael Patrick Cashin (1864–1926), a major political figure in Newfoundland and briefly prime minister of Newfoundland in 1919. Cashin was born in Cape Broyle, educated in St. Bonaventure's College, St. John's, and worked briefly for his father's firm (1908–11) before heading to western Canada to join a railway firm. He was in Fort William when war broke out and initially joined the Canadian Army but was persuaded by his father to return to Newfoundland. He signed up as a private with F Company, then recruiting for the Newfoundland Regiment. He was promoted to corporal following training and sent overseas in June 1915. He served with the depot at Ayr, Scotland, where he was commissioned on 15 October 1915. He joined the regiment in France in April 1916 but was wounded while on patrol and hospitalized at No. 3 General Hospital at Wandsworth, London. On recovery he was sent to Ayr. He was reassigned to the British Machine Gun Corps while retaining his commission in the Newfoundland Regiment and served in France with the British Third Army, eventually rising to the rank of Major in command of a machine gun company. He returned to Newfoundland where he rejoined his father's business. Cashin was a vocal critic of the Commission of Government and was a member of the National Convention sent to England to discuss Newfoundland's future forms of government. He was a vocal anti-Confederate and the leader of the Responsible Government group at the National Convention. Following confederation in 1949 he played an active role in provincial politics, eventually serving as Progressive Conservative Opposition leader from 1951 to 1953 when he resigned to run as an indepen-

dent in the Federal election. He retired from politics in 1953 and served as director of Civil Defence in Newfoundland until his retirement in the mid-1960s. He died in St. John's on 21 May, 1971. Cashin, *My Life and Times*, chapters 8–12 *passim;* Smallwood and Pitt, *Encyclopedia of Newfoundland and Labrador,* 1: 380–1.

21 James Cooper. Enlisted, 2 September 1914; BMEF, 20 August 1915; invalided to England, 21 October 1915; BEF, 28 March 1915; wounded, Beaumont Hamel, 1 July 1916; invalided to England; returned to BEF, 25 March 1917; Lance Corporal, 1 November 1917; Acting Corporal, 26 December 1917; embarked for Newfoundland on duty, 22 May 1918; discharged, St. John's, medically unfit, 7 December 1918. Cramm, *The First Five Hundred,* 165.

22 A "cushy" was a slight wound that took its recipient out of battle.

23 The horse probably belonged to regimental transport.

24 The trenches at Suvla were developed over time and so were deeper and wider. Up to this point those on the Western Front were not intended for long-term occupation but as a starting point for offensives against the German lines. British trenches conformed to the terrain and soil condition – if the ground was wet and stony the trench was shallow with a built-up area in front (parapet) and at the back (parados), usually constructed with sandbags. Trenches were normally dug deep enough to shelter a soldier and were narrow enough to make a difficult target for enemy artillery. They were kinked at intervals to diffuse blast, splinters, or shrapnel and to prevent an enemy from using enfilade fire against its occupants. Keegan, *The First World War,* 175–6.

25 Presumably referring to the regimental lock-up.

26 Private George Robert Curnew, killed 24 April 1916. Nicholson, *The Fighting Newfoundlander,* 550.

27 Lieutenant Wallace Ross was killed at Beaumont Hamel. Nicholson, *Fighting Newfoundlander,* 275.

28 Corporal Levi Bellows of Curling, Newfoundland; Regimental Number 903; later reduced to ranks by general court martial on 31 March 1917. Enlisted on 4 January 1915. Served with the regiment at Suvla and with the BEF in France. He was taken prisoner at the Battle of Monchy-le-Preux on 14 April 1917. Veterans Affairs Canada, The Royal Newfoundland Regiment, Regimental History Ledger Sheets, vol. 3.

29 According to James Steele, Bowers Hall, Barkisland, was the residence of friends of the Steele family, the Bottomleys. Bottomley was works manager at MacIntosh's, the confectionary factory in Scotland.

30 Stanley Charles Goodyear. Enlisted, 8 September 1914; appointed Lance Corporal, 3 October 1914; Corporal, 13 February 1915; Transport Sergeant, 14 June 1915; BMEF, 20 August 1915; served with the 1st Composite Battalion on the western Egyptian frontier, November 1915 to

February 1916; BEF, 2 March 1916; 2nd Lieutenant, 10 May 1916; Lieutenant, 1 August 1917; killed in action, Broembeek, 10 October 1917; awarded Military Cross, 28 December 1917. Cramm, *The First Five Hundred*, 187.

31 Nicholson states that both Greene and Lance Corporal Richard E. Hynes were awarded the Distinguished Conduct Medal on this occasion. However, in Appendix B Honours and Awards Nicholson lists only Hynes as a recipient of that award. Sergeant, later Company Sergeant Major Gregory L. Greene is recorded as receiving the Military Medal for putting enemy machineguns out of action at Kieberg Ridge. See Nicholson, *The Fighting Newfoundlander*, 245, 485, Appendix B.

32 Lieut. Colonel Hadow was awarded a CMG on this occasion. See Nicholson, *The Fighting Newfoundlander*, 244.

33 It is not clear from the context whether this is Gerald or Wilfred Ayre. Gerald W. Ayre was a grandson of C.R. Ayre, founder of one of St John's mercantile firms, and was related to Captain Eric Ayre and Second Lieutenant Wilfred D. Ayre, both officers with the regiment. He was killed at Beaumont Hamel. Nicholson, *The Fighting Newfoundlander*, 275.

34 Robert Stewart Errol Munn of St. John's, Newfoundland. Regimental Number 787; enlisted, 21 December 1914; wounded by shrapnel, 19 June 1916; sent to base hospital, 22 June 1916 and subsequently sent to Wandsworth hospital in England; returned to Newfoundland, August 1916; discharged. Veterans Affairs Canada, The Royal Newfoundland Regiment, Regimental History Ledger Sheets, vol. 2.

35 John Joseph Sheehan. Enlisted, 2 September 1914; BMEF, 20 August 1915; BEF, 14 March 1916; wounded Beaumont Hamel, 1 July 1916; invalided to England, 5 July 1916; Lance Corporal, 27 October 1916; returned to BEF, 30 December 1916; Corporal, 9 February 1917; wounded, Sailly-Saillisel, 24 February 1917; discharged, medically unfit, 8 December 1917; enlisted Newfoundland Forestry Battalion, 11 December 1917; Sergeant, 12 December 1917; embarked for United Kingdom, 21 December 1917; died of pneumonia, 28 December 1917. See Cramm, *The First Five Hundred*, 282.

36 Actually 33 *regiments*, each equivalent to a British brigade of three battalions of about 1000 each.

37 Stokes mortars, three-inch, later 81 mm, smoothbore infantry mortar capable of high angle fire from a trench. Griffith, *Battle Tactics of the Western Front*, 105.

38 Jim Steele in his account reported nineteen casualties, including the three officers in charge, two of whom, Lieutenants Strong and Greene, were sent to hospital. Steele, "A Few Notes."

39 Presumably a tetanus innoculation.

CHAPTER EIGHT

1 Steele, "A Few Notes."
2 Nicholson, *The Fighting Newfoundlander*, 248.
3 Gillon, *The 29th Division*, 79; Cheyne, *The Last Great Battle of the Somme*, 116–17.
4 Atwater, "Echoes and Voices," 210.
5 Archives, CNS, MUN, MF 147. Steele Diary, 337.
6 PANL, Major A. Raley, "Beaumont Hamel," 36.
7 Archives, CNS, MUN, MF 147. Steele Diary, 339.
8 Archives, CNS, MUN, MF 147. Steele Diary, 340.
9 "Everyone was happy and light-hearted, and the various battalions marched off whistling, singing and cheering ... everyone kept in very light spirits." Steele, "A Few Notes."
10 PANL, Raley, "Beaumont Hamel," 37.
11 Cheyne, *Beaumont Hamel 1916*, 117–18.
12 Gillon, *The 29th Division*, 80–1.
13 PANL, Raley, "Beaumont Hamel," 38.
14 PANL, Raley, "Beaumont Hamel," 40.
15 PANL, Aa-2-4 War Diaries, Entry 1 July 1916; Nicholson, *The Fighting Newfoundlander*, 270–2; Steele, "A Few Notes."
16 Stacey, *Memoirs of a Blue Puttee*, 85.
17 PANL, Aa-2-4, War Diaries, Entry 1 July 1916; Raley, "Beaumont Hamel," 43.
18 Steele, "A Few Notes."
19 Ronald Dunn. Regimental Number 1088; born in Bayley's Cove, Bonavista Bay, Newfoundland, son of Eli Dunn; joined the regiment on 8 February 1915; BMEF in Gallipoli; wounded at Beaumont Hamel, 1 July 1916; discharged, 18 January 1918. Veterans Affairs Canada, Royal Newfoundland Regiment.
20 This recollection appears in the film *Beaumont Hamel: A Battle Remembered* prepared by Educational Technology, Memorial University of Newfoundland [1991] to mark the seventy-fifth anniversary of the battle.
21 PANL Aa-2-4 War Diaries. Entry 1 July 1916; James R. Steele, "A Few Notes."
22 This is the figure given by Nicholson. David Parsons in his article "Stand To," based on an exhaustive search through the regiment's personnel records, gives the number of wounded as 438, of whom 29 died of wounds, killed in action as 133, and missing, presumed dead, as 139, bringing the total up to 710. If that is the case, then the regiment suffered as much as the 10th West Yorks. Nicholson, *The Fighting Newfoundlander*, 274; Parsons, "Stand To," 16–17.

23 Nicholson, *The Fighting Newfoundlander*, 274–5 n.
24 NAC, CBC, Interview of Col. A.L. Hadow, May 1964.
25 Nicholson, *The Fighting Newfoundlander*, 275–6; Anthony J. Stacey
    Diary; PANL, Aa-2-4 War Diaries, Entry 7 July 1916; NAC 150/Accession
    1992–93/171, Box 5 O-89; letter James Forbes Robertson to Mr.
    [Samuel] Steele, 27 August 1916; NAC 150/Accession 1992–93/171, Box
    5 File 0-89; James Forbes Robertson to S.O. Steele, 27 August 1916.

CHAPTER NINE

 1 *Daily News*, 3 July 1916.
 2 *Daily News*, 7 July 1916.
 3 Kerr and Holdsworth, eds, *A Historical Atlas of Canada*. 3: plate 26.
 4 "The Thoughts of Theobold," *Daily News*, 8 July 1916.
 5 NAC, 150/Accession 1992–93/171, Box 5, File 0-89, S.O. Steele to Capt.
    Timewell, 13 July 1916.
 6 Ibid., S.O. Steele to Sir W.E. Davidson, 14 July 1916; W.E. Davidson to
    S.O. Steele, 14 July 1916.
 7 Ibid., W. E. Davidson to Capt. A. Timewell, 18 July 1916.
 8 Ibid., Samuel O. Steele to Sir Walter Davidson, 18 December 1916.
 9 PANL, MG 632, file #34, Censorship in Newfoundland.
10 Ibid.
11 NAC, 150/Accession 1992–93/171, Box 5, File 0-89, W.E. Davidson to
    S.O. Steele, 29 March 1917.
12 Nicholson, *The Fighting Newfoundlander*, 506–7.
13 Ibid., 514–16.
14 PANL, "Reunion Daily News, 5 October 1921," *The Veteran* 1 no. 4
    (December 1921): 72.

Bibliography

PRIMARY SOURCES

NATIONAL ARCHIVES OF CANADA

Canadian Broadcasting Corporation, Radio: Flanders Fields ISN 117216.
  CO7721(3) Flanders Fields: [background interview] Hadow, A.J. – Royal
  Newfoundland Regiment – Interview May 1964.
150/Accession 1992–93/171, box 5.
MG31, G19, Nicholson, Gerald William Lingen. Correspondence 1903–1979.
  Vol 1. Correspondence 1961–1963.

PROVINCIAL ARCHIVES OF NEWFOUNDLAND AND LABRADOR

Aa-2-4 War Diaries, Royal Newfoundland Regiment
MG 9, Captain C. Sydney Frost Collection
MG 632, file no. 34, "Censorship in Newfoundland"
The Veteran

CENTRE FOR NEWFOUNDLAND STUDIES,
MEMORIAL UNIVERSITY OF NEWFOUNDLAND

Guide for Officers, Warrant Officers, Non-Commissioned Officers,
and Privates of the Church Lads' Brigades Infantry Training
1910

CENTRE FOR NEWFOUNDLAND STUDIES ARCHIVES,
MEMORIAL UNIVERSITY OF NEWFOUNDLAND

MF 147, Owen W. Steele, "Diary of Newfoundland Regiment"
Newfoundland Patriotic Association, letters and statistics, 1917

PRIVATE COLLECTION, JAMES STEELE

James R. Steele, "A Few Notes Relative to the Newfoundland Regiment and
the Somme Battle, July 1st, 1916," forwarded to Lt Col Rendell, Department of Militia, 25 June 1920

VETERANS AFFAIRS CANADA

The Royal Newfoundland Regiment, Regimental History Ledger Sheets. Original in the Provincial Archives of Newfoundland and Labrador.

PRINTED PRIMARY SOURCES

Who's Who In and From Newfoundland, 1927. St. John's, 1927.

NEWSPAPERS

Daily News
The Evening Telegram

SECONDARY SOURCES

Atwater, James D. "Echoes and Voices Summoned from a Half-Hour in
Hell," *Smithsonian* (November 1987)
Brown, Elizabeth. "Citizen Soldiers: Political Pawns: The Newfoundland Vol-
unteer Companies 1860-1870." History Honours Thesis, Memorial Uni-
versity of Newfoundland, 1998
Cashin, Major Peter. *My Life and Times 1890-1919.* R.E. Buehler, ed. St.
John's: Breakwater Books, 1976

Cheyne, G.Y. *The Last Battle of the Somme: Beaumont Hamel 1916*. Edinburgh: John Donald Publishers, 1988

Churchill, Jason. "'No Surrender until Prussian Militarism is Crushed': The Loyal Orange Association in Newfoundland and the First World War effort." History Honours Thesis, Memorial University of Newfoundland, 1996

Cramm, Richard. *The First Five Hundred, being a Historical Sketch of the Military Operations of the Royal Newfoundland Regiment in Gallipoli and on the Western Front during the Great War, (1914–1918) together with the Individual Military Records and Photographs Where Obtainable of the Men of the First Contingent, Known as "The First Five Hundred," or "The Blue Puttees."* Albany, New York: C.F. Williams, 1921

Dancocks, Daniel G. *Welcome to Flander's Fields: The First Canadian Battle of the Great War: Ypres, 1915*. Toronto: McClelland & Stewart Inc., 1989

Dictionary of Canadian Biography. Vol. 14. Toronto: University of Toronto Press, 1998

The Dictionary of National Biography: The Concise Dictionary. Part II, 1901–1950. Oxford: University Press, 1961

The Dissenting Church of Christ in St. John's, 1775–1975. St. John's: St. David's Presbyterian Church, 1975

Facey-Crowther, David R. "Militiamen and Volunteers: The New Brunswick Militia 1787–1871" *Acadiensis* (Autumn 1990): 148–73

Fussell, Paul. *The Great War and Modern Memory*. London: Oxford University Press, 1975

Gallishaw, John. *Trenching at Gallipoli: The Personal Narrative of a Newfoundlander with the Ill-Fated Dardanelles Expedition*. New York: A.L. Burt, 1916

Gilbert, William. "Newfoundland Regiment: Its Volunteer Patterns, August 1914 to July 17, 1917." History 3120 paper, Maritime History Group, Memorial University of Newfoundland, n.d.

Gillon, Stair, *The Story of the 29th Division: A Record of Gallant Deeds*. London: Thomas Nelson and Sons, 1925

Griffith, Paddy. *Battle Tactics of the Western Front: The British Army's Art of Attack 1916–18*. New Haven and London: Yale University Press, 1994

Harding, Bob. "Wireless in Newfoundland, 1901–1933." *Newfoundland Quarterly* (Summer/Fall 1901): 15–17

Haythornthwaite, Philip J. *Gallipoli 1915: Frontal Assault on Turkey*. London: Osprey Publishing, 1991

Hynes, Samuel. *The Soldier's Tale: Bearing Witness to Modern War*. Toronto: Penguin Books, 1998

James, Robert Rhodes. *Gallipoli*. London: Papermac, 1989

Keegan, John. *The Face of Battle*. Harmondsworth: Penguin, 1978

– *The First World War*. Toronto: Key Porter, 1998

Kerr, D.P. and D.W. Holdsworth, eds. *A Historical Atlas of Canada*. Vol. III. Toronto: University of Toronto Press, 1990

Lind, Mayo. *The Letters of Mayo Lind: Newfoundland's Unofficial War Correspondent 1914–1916*. St. John's, Creative Publishers, 2001

Lucas, Sir Charles. *The Empire at War*, Vol II, London: Oxford University Press, 1923

McDonald, Ian D.H. *"To Each His Own": William Coaker and the Fisherman's Protective Union in Newfoundland Politics, 1908–1925*. J.K. Hiller, ed. St. John's: Institute of Social and Economic Research, Memorial University of Newfoundland, 1987

Macdonald, Lyn. *1914-1918. Voices and Images of the Great War*. London: Penguin, 1991

– *Somme*. New York: Penguin, 1993

Macfarlane, David. *The Danger Tree: Memory, War and the Search for a Family's Past*. Toronto: Macfarlane, Walter and Ross, 1991

Middlebrook, Martin. *The First Day on the Somme: 1 July 1916*. Harmondsworth: Penguin, 1984

Moorehead, Alan. *Gallipoli*. Hertfordshire: Wordsworth Editions, 1997

Murray, Chris, ed. *The Hutchinson Dictionary of the Arts*. Oxford, Helicon, c. 1994

Nicholson, G.W. L. *More Fighting Newfoundlanders*. St. John's: Government of Newfoundland, 1969

– *The Fighting Newfoundlander: A History of the Royal Newfoundland Regiment*. St. John's: Government of Newfoundland, 1964

Noel, S.J.R. *Politics in Newfoundland*. Toronto: University of Toronto Press, 1971

O'Brien, Patricia. "The Newfoundland Patriotic Association: The Administration of the War Effort, 1914–1918." Master's thesis, Memorial University of Newfoundland, 1981

O'Neill, Paul. *The Oldest City: The Story of St. John's, Newfoundland*. Erin, Ontario: Press Porcepic, 1975

– *A Seaport Legacy: The Story of St. John's, Newfoundland*. Erin, Ontario: Press Porcepic, 1976

Parsons, Andrew. "Morale and Cohesion in the Royal Newfoundland Regiment 1914-18" Master's thesis, Memorial University of Newfoundland, 1995

Parsons, W.D. "Stand To," *Journal of the Western Front Association* 22 (April 1988): 16–17

– *Pilgrimage: A Guide to the Royal Newfoundland Regiment in World War One*. St. John's: Creative Publishers, 1994.

Penney, Shane. "The Church Lads Brigade and Catholic Cadet Corps in St. John's as Examples of the Transference of British Youth Culture Prior to

1914". History Honours thesis, Memorial University of Newfoundland, 2000

Poole, Cyril F. and Robert H. Cuff, eds. *Encyclopedia of Newfoundland and Labrador*. Vol. 5. St John's: Harry Cuff Publications 1991–94

Pritchett, Sterling D. *A History of the Church Lads' Brigade in Newfoundland. 100th Anniversary*. St. John's: Creative Publishers, 1991

Ryan, Shannon. *The Ice Hunters: A History of Newfoundland Sealing to 1914*. St. John's: Breakwater Press, 1994

– "Newfoundland Spring Sealing Disasters to 1914." *The Northern Mariner/Le Marin du Nord* 3, no. 3 (July/Juillet 1993): 40–4

St. John's City Directory 1919. Halifax: Regal Print & Litho Limited, 1919

Sharpe, Christopher A. "The 'Race of Honour': An Analysis of Enlistments and Casualties in the Armed Forces of Newfoundland: 1914–1918," *Newfoundland Studies* 4, 1 (1988): 27–55

Smallwood, J.R., and Robert D.W. Pitt, eds. *Encyclopedia of Newfoundland and Labrador*. Vols. 1, 2, 3 and 5. (St. John's: Harry Cuff Publications, 1991–94)

Stacey, A.J. and Jean Edwards, *Memoirs of a Blue Puttee: The Newfoundland Regiment in World War One*. St. John's: DRC Publishers, 2002.

Tait, R.H. "With the Regiment at Gallipoli," PANL, *The Veteran* 1, no. 2 (April 1921)

Terraine, John. *The First World War 1914–18*. London: Papermac, 1983

Warren, Denise Gale. "The Patriotic Association of the Women of Newfoundland: 1914–1918." History Honours thesis, Memorial University of Newfoundland, 1996

Who Was Who 1961–1970, Vol 6. London: Adam and Charles Black, 1970

Ziegler, Philip. *Omdurman*. London: Collins, 1973

INTERNET SOURCES

Dictionary of Newfoundland English, www.heritage.nf.ca/dictionary/d8ction.html

Encarta Encyclopedia Online Deluxe

Index